SCHAUM'S OUTLINE OF

THEORY AND PROBLEMS

of

ENGINEERING ECONOMICS

•

by

JOSÉ A. SEPULVEDA, Ph.D.
Associate Professor of Industrial Engineering
University of Central Florida

WILLIAM E. SOUDER, Ph.D.
Professor of Industrial Engineering
University of Pittsburgh

and

BYRON S. GOTTFRIED, Ph.D.
Professor of Industrial Engineering
Engineering Management and Operations Research
University of Pittsburgh

SCHAUM'S OUTLINE SERIES

McGRAW-HILL, INC.

New York St. Louis San Francisco Auckland Bogotá Caracas
Hamburg Lisbon London Madrid Mexico Milan Montreal
New Delhi Paris San Juan São Paulo Singapore
Sydney Tokyo Toronto

JOSÉ A. SEPULVEDA is an Associate Professor of Industrial Engineering at the University of Central Florida. He holds a Ph.D. in Industrial Engineering and an M.P.H. from the University of Pittsburgh. His research and teaching interests are in health operations research and economic feasibility analysis. He is a Registered Professional Engineer.

WILLIAM E. SOUDER is a Professor of Industrial Engineering at the University of Pittsburgh. Since 1972, he has been at the University of Pittsburgh, where he teaches courses in Engineering Management and Behavioral Systems and directs the Technology Management Studies Research Group. He is the author of two other books and over one hundred technical papers.

BYRON S. GOTTFRIED is a Professor of Industrial Engineering, Engineering Management, and Operations Research at the University of Pittsburgh. He received his Ph.D. from Case-Western Reserve University (1962), and has been a member of the Pitt faculty since 1970. His primary interests are in the development of complex technical and business applications of computers. Dr. Gottfried is the author of several textbooks, as well as *Introduction to Engineering Calculations* in the Schaum's Outline Series.

Schaum's Outline of Theory and Problems of
ENGINEERING ECONOMICS

5 6 7 8 9 10 11 12 13 14 15 SH SH 9 8 7 6 5 4 3 2 1 0

ISBN 0-07-023834-0

Sponsoring Editor, David Beckwith
Editing Supervisor, Marthe Grice
Production Manager, Nick Monti

Library of Congress Cataloging in Publication Data

Sepulveda, José A.
Schaum's outline of theory and problems of engineering economics.

(Schaum's outline series)
Includes index.
1. Engineering economy. I. Souder, William E.
II. Gottfried, Byron S., 1934– III. Title.
TA177.4.S47 1984 658.1'55 84-778
ISBN 0-07-023834-0

Cover design by Amy E. Becker.

Preface

Despite remarkable technological advances during the past several decades, most major engineering decisions are based on economic considerations—a situation that is unlikely to change in the years ahead. Hence the importance of economic principles to all undergraduate engineering students, regardless of their particular disciplinary interests.

This Schaum's Outline contains a clear and concise review of the principles of engineering economics, together with a large number of solved problems. Most chapters also contain a list of supplementary problems, which the readers may solve themselves. Thus, readers receive an exposure to the theory, as well as an opportunity to become actively involved in the application of this theory to typical (though simple) problem situations.

The book is designed to complement a standard undergraduate course in engineering economics. The first five chapters consider the mathematics of compound interest, emphasizing the time value of money. Chapters 6 through 9 discuss the application of this material in various decision-making criteria, and Chapter 10 deals with equipment replacement and retirement decisions. Chapter 11 considers the important topics of depreciation and taxes, and their impact on the decision-making process. Finally, Chapter 12 presents a realistic economic feasibility study.

The four appendixes to the book contain tables of various compound interest factors. Such tables continue to be useful, even in an era of electronic calculators and personal computers.

JOSÉ A. SEPULVEDA
WILLIAM E. SOUDER
BYRON S. GOTTFRIED

Contents

CONTENTS

Chapter 1

Basic Concepts

1.1 INTEREST

Interest is a fee that is charged for the use of someone else's money. The size of the fee will depend upon the total amount of money borrowed and the length of time over which it is borrowed.

Example 1.1 An engineer wishes to borrow \$20 000 in order to start his own business. A bank will lend him the money provided he agrees to repay \$920 per month for two years. How much interest is he being charged?

The total amount of money that will be paid to the bank is $24 \times \$920 = \$22\,080$. Since the original loan is only \$20 000, the amount of interest is $\$22\,080 - \$20\,000 = \$2080$.

Whenever money is borrowed or invested, one party acts as the lender and another party as the borrower. The lender is the owner of the money, and the borrower pays interest to the lender for the use of the lender's money. For example, when money is deposited in a savings account, the depositor is the lender and the bank is the borrower. The bank therefore pays interest for the use of the depositor's money. (The bank will then assume the role of the lender, by loaning this money to another borrower, at a higher interest rate.)

1.2 INTEREST RATE

If a given amount of money is borrowed for a specified period of time (typically, one year), a certain percentage of the money is charged as interest. This percentage is called the *interest rate.*

Example 1.2 (*a*) A student deposits \$1000 in a savings account that pays interest at the rate of 6% per year. How much money will the student have after one year? (*b*) An investor makes a loan of \$5000, to be repaid in one lump sum at the end of one year. What annual interest rate corresponds to a lump-sum payment of \$5425?

(*a*) The student will have his original \$1000, plus an interest payment of $0.06 \times \$1000 = \$60$. Thus, the student will have accumulated a total of \$1060 after one year. (Notice that the interest rate is expressed as a decimal when carrying out the calculation.)

(*b*) The total amount of interest paid is $\$5425 - \$5000 = \$425$. Hence the annual interest rate is

$$\frac{\$425}{\$5000} \times 100\% = 8.5\%$$

Interest rates are usually influenced by the prevailing economic conditions, as well as the degree of risk associated with each particular loan.

1.3 SIMPLE INTEREST

Simple interest is defined as a fixed percentage of the *principal* (the amount of money borrowed), multiplied by the life of the loan. Thus,

$$I = niP \tag{1.1}$$

where $I \equiv$ total amount of simple interest
$n \equiv$ life of the loan
$i \equiv$ interest rate (expressed as a decimal)
$P \equiv$ principal

It is understood that n and i refer to the same unit of time (e.g., the year).

Normally, when a simple interest loan is made, nothing is repaid until the end of the loan period; then, both the principal and the accumulated interest are repaid. The total amount due can be expressed as

$$F = P + I = P(1 + ni) \tag{1.2}$$

Example 1.3 A student borrows \$3000 from his uncle in order to finish school. His uncle agrees to charge him simple interest at the rate of $5\frac{1}{2}\%$ per year. Suppose the student waits two years and then repays the entire loan. How much will he have to repay?

By (*1.2*), $F = \$3000[1 + (2)(0.055)] = \$3330$.

1.4 COMPOUND INTEREST

When interest is *compounded*, the total time period is subdivided into several *interest periods* (e.g., one year, three months, one month). Interest is credited at the end of each interest period, and is allowed to accumulate from one interest period to the next.

During a given interest period, the current interest is determined as a percentage of the total amount owed (i.e., the principal plus the previously accumulated interest). Thus, for the first interest period, the interest is determined as

$$I_1 = iP$$

and the total amount accumulated is

$$F_1 = P + I_1 = P + iP = P(1 + i)$$

For the second interest period, the interest is determined as

$$I_2 = iF_1 = i(1 + i)P$$

and the total amount accumulated is

$$F_2 = P + I_1 + I_2 = P + iP + i(1 + i)P = P(1 + i)^2$$

For the third interest period,

$$I_3 = i(1 + i)^2P \qquad F_3 = P(1 + i)^3$$

and so on. In general, if there are n interest periods, we have (dropping the subscript):

$$F = P(1 + i)^n \tag{1.3}$$

which is the so-called *law of compound interest.* Notice that F, the total amount of money accumulated, increases exponentially with n, the time measured in interest periods.

Example 1.4 A student deposits \$1000 in a savings account that pays interest at the rate of 6% per year, compounded annually. If all of the money is allowed to accumulate, how much will the student have after 12 years? Compare this with the amount that would have accumulated if simple interest had been paid.

By (*1.3*),

$$F = \$1000(1 + 0.06)^{12} = \$2012.20$$

Thus, the student's original investment will have more than doubled over the 12-year period.

If simple interest had been paid, the total amount that would have accumulated is determined by (*1.2*) as

$$F = \$1000[1 + (12)(0.06)] = \$1720.00$$

1.5 THE TIME VALUE OF MONEY

Since money has the ability to earn interest, its value increases with time. For instance, \$100 today is equivalent to

$$F = \$100(1 + 0.07)^5 = \$140.26$$

five years from now if the interest rate is 7% per year, compounded annually. We say that the *future worth* of \$100 is \$140.26 if $i = 7\%$ (per year) and $n = 5$ (years).

Since money increases in value as we move from the present to the future, it must decrease in value as we move from the future to the present. Thus, the *present worth* of \$140.26 is \$100 if $i = 7\%$ (per year) and $n = 5$ (years).

Example 1.5 A student who will inherit \$5000 in three years has a savings account that pays $5\frac{1}{2}\%$ per year, compounded annually. What is the present worth of the student's inheritance?

Equation (*1.3*) may be solved for P, given the value of F:

$$P = \frac{F}{(1+i)^n} = \frac{\$5000}{(1+0.055)^3} = \$4258.07$$

The present worth of \$5000 is \$4258.07 if $i = 5\frac{1}{2}\%$, compounded annually, and $n = 3$.

1.6 INFLATION

National economies frequently experience *inflation*, in which the cost of goods and services increases from one year to the next. Normally, inflationary increases are expressed in terms of percentages which are compounded annually. Thus, if the present cost of a commodity is PC, its future cost, FC, will be

$$\text{FC} = \text{PC}(1+\lambda)^n \tag{1.4}$$

where $\lambda \equiv$ annual inflation rate (expressed as a decimal)
$n \equiv$ number of years

Example 1.6 An economy is experiencing inflation at the rate of 6% per year. An item presently costs \$100. If the 6% inflation rate continues, what will be the price of this item in five years?

By (*1.4*), $\text{FC} = \$100(1+0.06)^5 = \$133.82$.

In an inflationary economy, the value (buying power) of money decreases as costs increase. Thus,

$$\frac{F}{P} = \frac{\text{PC}}{\text{FC}} = \frac{1}{(1+\lambda)^n}$$

or

$$F = \frac{P}{(1+\lambda)^n} \tag{1.5}$$

where F is the future worth, measured in today's dollars, of a present amount P.

Example 1.7 An economy is experiencing inflation at an annual rate of 6%. If this continues, what will \$100 be worth five years from now, in terms of today's dollars?

From (*1.5*),

$$F = \frac{\$100}{(1+0.06)^5} = \$74.73$$

Thus \$100 in five years will be worth only \$74.73 in terms of today's dollars. Stated differently, in five years \$100 will be required to purchase the same commodity that can now be purchased for \$74.73.

If interest is being compounded at the same time that inflation is occurring, then the future worth can be determined by combining (*1.3*) and (*1.5*):

$$F = \frac{P(1+i)^n}{(1+\lambda)^n} = P\left(\frac{1+i}{1+\lambda}\right)^n$$

or, defining the *composite interest rate*,

$$\theta \equiv \frac{i-\lambda}{1+\lambda} \tag{1.6}$$

we have

$$F = P(1+\theta)^n \tag{1.7}$$

Observe that θ may be negative.

Example 1.7 An engineer has received \$10 000 from his employer for a patent disclosure. He has decided to invest the money in a 15-year savings certificate that pays 8% per year, compounded annually. What will be the final value of his investment, in terms of today's dollars, if inflation continues at the rate of 6% per year?

A composite interest rate can be determined from (*1.6*):

$$\theta = \frac{0.08-0.06}{1+0.06} = 0.0189$$

Substituting this value into (*1.7*), we obtain

$$F = \$10\,000(1+0.0189)^{15} = \$13\,242.61$$

(If more significant figures are included in the value for θ, the future value \$13 236.35 is obtained.)

1.7 TAXES

In most situations, the interest that is received from an investment will be subject to taxation. Suppose that the interest is taxed at a rate t, and that the period of taxation is the same as the interest period (e.g., one year). Then the tax for each period will be $T = tiP$, so that the net return to the investor (after taxes) will be

$$I' = I - T = (1-t)iP \tag{1.8}$$

If the effects of taxation and inflation are both included in a compound interest calculation, (*1.7*) may still be used to relate present and future values, provided the composite interest rate is redefined as

$$\theta \equiv \frac{(1-t)i-\lambda}{1+\lambda} \tag{1.9}$$

Example 1.8 Refer to Example 1.7. Suppose the engineer is in the 32% tax bracket, and is likely to remain there throughout the lifetime of the certificate. If inflation continues at the rate of 6% per year, what will be the value of his investment, in terms of today's dollars, when the certificate matures?

Let us assume that the engineer is able to invest the entire \$10 000 in a savings certificate and that the 32% tax bracket includes all federal, state, and local taxes. By (*1.9*),

$$\theta = \frac{(1-0.32)(0.08)-0.06}{1+0.06} = -0.00528$$

and (*1.7*) then gives

$$F = \$10\,000(1-0.00528)^{15} = \$9236.61$$

Because of the combined effects of inflation and taxation, θ is negative, and the engineer ends up with *less real purchasing power* after 15 years than he has today. (To make matters worse, the engineer will most likely have to pay taxes on the original \$10 000, substantially reducing the amount of money available for investment.)

The subject of taxation is considered in much greater detail in Chapter 11.

1.8 CASH FLOWS

A *cash flow* is the difference between total cash receipts (*inflows*) and total cash disbursements (*outflows*) for a given period of time (typically, one year). Cash flows are very important in engineering economics because they form the basis for evaluating projects, equipment, and investment alternatives.

The easiest way to visualize a cash flow is through a *cash flow diagram*, in which the individual cash flows are represented as vertical arrows along a horizontal time scale. Positive cash flows (net inflows) are represented by upward-pointing arrows, and negative cash flows (net outflows) by downward-pointing arrows; the length of an arrow is proportional to the magnitude of the corresponding cash flow. Each cash flow is assumed to occur at the *end* of the respective time period.

Example 1.9 A company plans to invest $500 000 to manufacture a new product. The sale of this product is expected to provide a net income of $70 000 a year for 10 years, beginning at the end of the first year. Figure 1-1 is the cash flow diagram for this proposed project. Notice that the initial $500 000 investment is represented by a downward-pointing arrow located at the end of year 0 (i.e., at the beginning of year 1). Each annual net income ($70 000) is indicated by an upward-pointing arrow located at the end of the corresponding year.

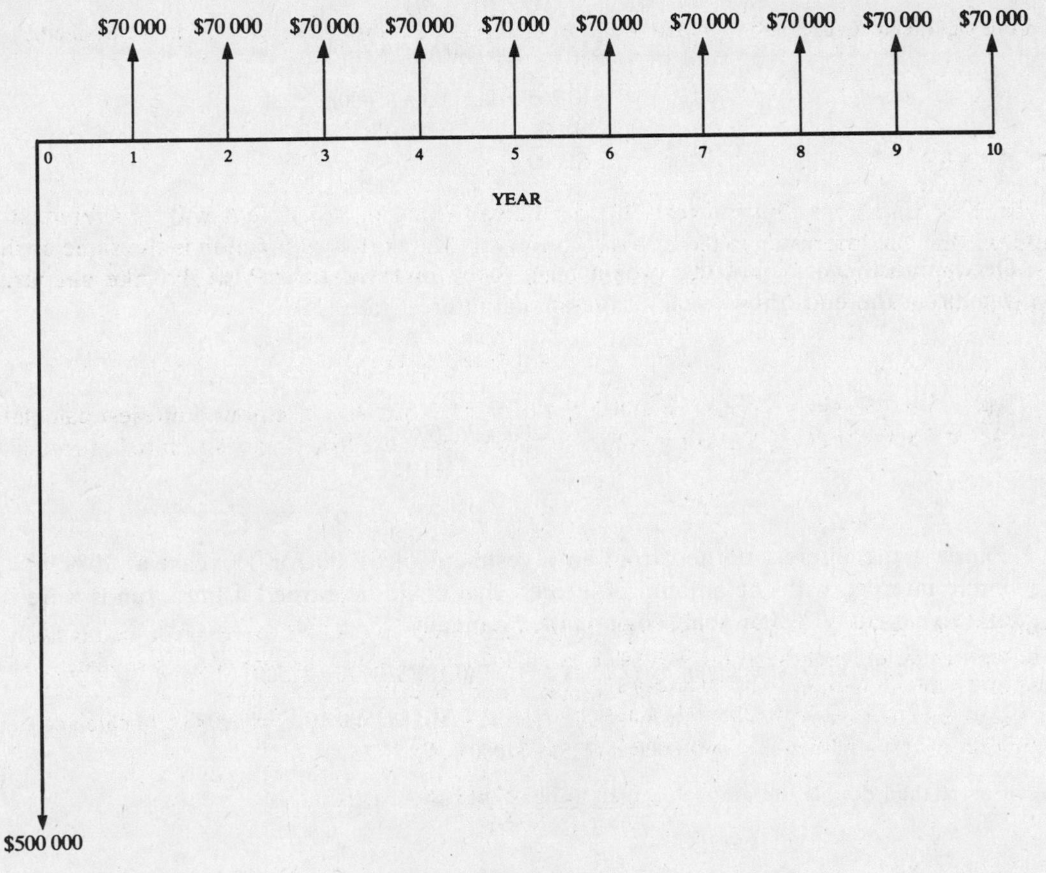

Fig. 1-1

In a lender-borrower situation, an inflow for the one is an outflow for the other. Hence, the cash flow diagram for the lender will be the mirror image in the time line of the cash flow diagram for the borrower.

Solved Problems

1.1 The ABC Company deposited \$100 000 in a bank account on June 15 and withdrew a total of \$115 000 exactly one year later. Compute: (*a*) the *interest* which the ABC Company received from the \$100 000 investment, and (*b*) the annual *interest rate* which the ABC Company was paid.

(*a*)
$$I = \$115\,000 - \$100\,000 = \$15\,000$$

(*b*)
$$i = \frac{\$15\,000/\text{year}}{\$100\,000} \times 100\% = 15\% \text{ per year}$$

1.2 What is the annual rate of simple interest if \$265 is earned in four months on an investment of \$15 000?

From (*1.2*),

$$F = P(1 + ni)$$
$$\$15\,265 = \$15\,000\left(1 + \frac{4}{12}i\right)$$
$$\$15\,265 = \$15\,000 + \$5000i$$
$$\frac{\$265}{\$5000} = i$$
$$i = 5.3\%$$

1.3 Determine the principal that would have to be invested to provide \$200 of simple interest income at the end of two years if the annual interest rate is 9%

By (*1.1*),

$$I = niP$$
$$\$200 = 2(0.09)P$$
$$P = \$1111.11$$

1.4 Compare the interest earned from an investment of \$1000 for 15 years at 10% per annum simple interest, with the amount of interest that could be earned if these funds were invested for 15 years at 10% per year, compounded annually.

The simple interest is given by (*1.1*) as $I = (15)(0.10)(\$1000) = \$1500$. From (*1.3*),

$$I = F - P = P(1 + i)^n - P = \$1000(1 + 0.10)^{15} - \$1000$$
$$= \$1000(4.17725) - \$1000 = \$3177.25$$

or more than double the amount earned using simple interest.

1.5 At what annual interest rate is \$500 one year ago equivalent to \$600 today?

From (*1.3*),

$$\$600 = \$500(1 + i)^1 \qquad \text{or} \qquad i = 20\%$$

1.6 Suppose that the interest rate is 10% per year, compounded annually. What is the minimum amount of money that would have to be invested for a two-year period in order to earn \$300 in interest?

From (*1.3*),

$$P = \frac{F}{(1+i)^n}$$

$$P = \frac{P + \$300}{(1+0.10)^2}$$

$$1.21\,P = P + \$300$$

$$P = \$1428.57$$

1.7 How long would it take for an investor to double his money at 10% interest per year, compounded annually?

By (*1.3*),

$$2P = P(1+0.10)^n$$

$$2 = (1.10)^n$$

$$n = \frac{\log 2}{\log 1.10} = 7.27 \text{ years}$$

Actually, since the interest is compounded only at the end of each year, the investor would have to wait 8 years.

1.8 Suppose that a man lends $1000 for four years at 12% per year simple interest. At the end of the four years, he invests the entire amount which he then has for 10 years at 8% interest per year, compounded annually. How much money will he have at the end of the 14-year period?

From (*1.2*) and (*1.3*),

$$F = P(1 + n_1 i_1)(1 + i_2)^{n_2} = \$1000[1 + (4)(0.12)](1 + 0.08)^{10}$$
$$= \$1000(1.48)(2.15892) = \$3195.21$$

1.9 Let the inflation rate be 6% per year. If a person deposits $50 000 in a bank account at 9% per annum simple interest for 10 years, will this effectively protect the purchasing power of the original principal?

The answer is not obvious, for the inflation rate, though the smaller, is compounded. From (*1.2*), the principal will grow to:

$$F = \$50\,000[1 + (10)(0.09)] = \$95\,000$$

By (*1.5*), the inflation will reduce the "real" purchasing power of these funds to

$$F = \frac{\$95\,000}{(1+0.06)^{10}} = \$53\,047.50 > \$50\,000$$

Thus, the purchasing power of the investment will be protected, and a small amount of interest will be earned.

1.10 Rework Problem 1.9, which is changed as follows: the individual is in the 35% tax bracket and pays taxes on all the interest received; the $50 000 is invested at 9% per year, compounded annually.

Now (*1.9*) gives

$$\theta = \frac{(1-0.35)(0.09) - 0.06}{1 + 0.06} = \frac{-0.00150}{1.06} < 0$$

With θ negative, (*1.7*) implies $F < P$: the investment is not protected, and there will be a (small) loss.

1.11 An individual wants to have \$2000 at the end of three years. How much would the individual have to invest at a 10% per year interest rate, compounded annually, in order to obtain a net of \$2000 after paying a \$250 early withdrawal fee at the end of the third year? Draw a cash flow diagram for the individual.

By *(1.3)*,

$$\$2250 = P(1+0.10)^3 \qquad \text{or} \qquad P = \$1690.46$$

The cash flow diagram for the individual is given in Fig. 1-2.

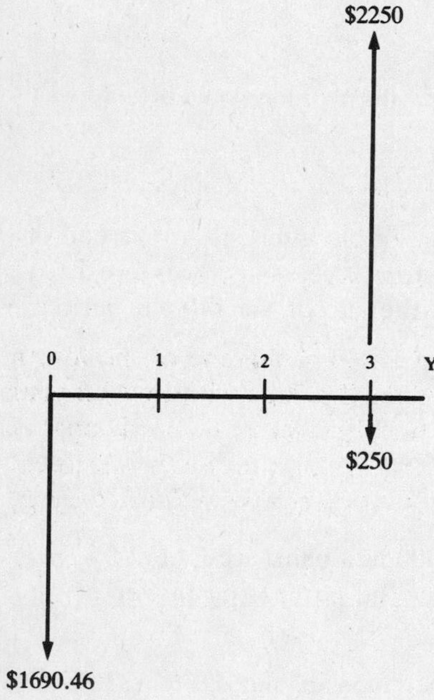

$2250

0 1 2 3 YEARS

$250

$1690.46

Fig. 1-2

Table 1-1

End of Year	Savings	Withdrawals	Cash Flows
0	\$600	\$ 0	−\$600
1		300	+300
2	500	300	−200
3		300	+300
4	500		−500
5		350	+350
6	400		−400
7	400	350	− 50
8	400		−400
9	400	350	− 50
10	400		−400

1.12 Suppose that you have a savings plan covering the next ten years, according to which you put aside \$600 today, \$500 at the end of every other year for the next five years, and \$400 at the end of each year for the remaining five years. As part of this plan, you expect to withdraw \$300 at the end of every year for the first 3 years, and \$350 at the end of every other year thereafter. (*a*) Tabulate your cash flows. (*b*) Draw your cash flow diagram.

(*a*) See Table 1-1.

(*b*) See Fig. 1-3.

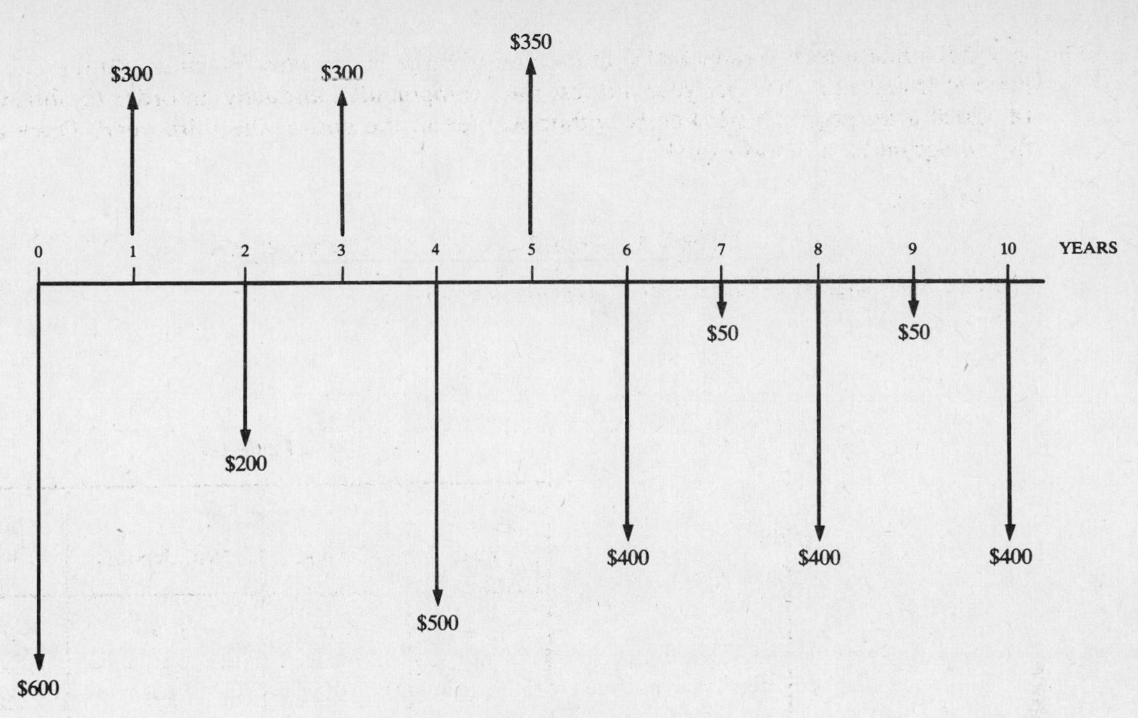

Fig. 1-3

1.13 Under your six-year savings plan, you deposit $1000 now, and $1000 at the end of the fourth year, in a bank account that earns 8% per year, compounded annually. You withdraw all your accumulated interest at the end of the second year, and the further interest plus principal at the end of the sixth year. (*a*) Tabulate the cash flows and the balance in your investment account. (*b*) Draw a cash flow diagram *for the bank*. (*c*) Compute the penalty (suffered by you) for the early withdrawal of interest at the end of the second year.

(*a*) See Table 1-2.

Table 1-2

End of Year	Deposits	Withdrawals	Account Balance for First $1000	Account Balance for Second $1000	Account Balance Total	Cash Flows
0	$1000	$ 0	$1000.00	---	$1000.00	−$1000.00
1	0	0	1080.00	---	1080.00	0
2	0	166.40	1000.00	---	1000.00	+166.40
3	0	0	1080.00	---	1080.00	0
4	1000	0	1166.40	$1000.00	2166.40	− 1000.00
5	0	0	1259.71	1080.00	2339.71	0
6	0	2526.88	1360.48	1166.40	2526.88	+ 2526.88
					NET CASH FLOW	+$693.28

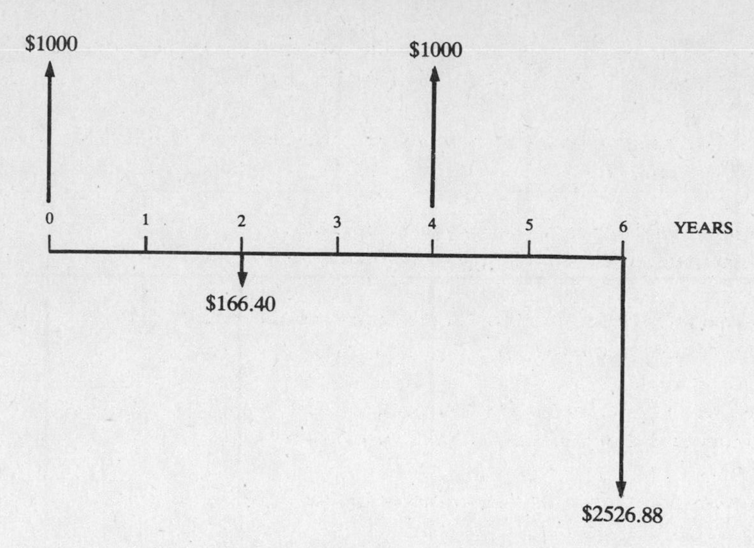

Fig. 1-4

(*b*) See Fig. 1-4.

(*c*) If the accumulated interest had not been withdrawn at the end of year 2, the $1000 invested at the start of the plan would have grown to

$$F = P(1+i)^n = \$1000(1.08)^6 = \$1586.87$$

and the total available for withdrawal at the end of year 6 would have been

$$\$1586.87 + \$1166.40 = \$2753.27$$

Hence, the net cash flow would have been $753.27, or $59.99 more.

Supplementary Problems

1.14 How much interest would be due at the end of one year on a loan of $10 000 if the interest is 12% per year? *Ans.* $1200

1.15 What is the annual interest rate on a $1000 loan in which all interest is paid at the end of the year, and a total of $1125 must be repaid at the end of the year? *Ans.* 12.5%

1.16 If $300 is earned in three months on an investment of $12 000, what is the annual rate of simple interest? *Ans.* 10%

1.17 How long will it take for an investment of $5000 to grow to $7500, if it earns 10% simple interest per year? *Ans.* 5 years

1.18 Find the principal of a loan in which the interest rate is $1\frac{1}{2}$% per month, payable monthly, and in which the borrower has just made the first monthly interest payment of $50. *Ans.* $3333.33

1.19 Find the principal, if the principal plus interest at the end of one and one-half years is $3360 for a simple interest rate of 8% per annum. *Ans.* $3000

1.20 Which is more desirable: investing $2000 at 6% per year compound interest for three years, or investing $2000 at 7% per year simple interest for three years? *Ans.* Simple interest is superior by $38.

1.21 At what rate of interest, compounded annually, will an investment triple itself in (*a*) 8 years? (*b*) 10 years? (*c*) 12 years? *Ans.* (*a*) 14.7%; (*b*) 11.6%; (*c*) 9.6%

1.22 What rate of interest, compounded annually, will result in the receipt of $15 938.48 if $10 000 is invested for 8 years? *Ans.* 6%

1.23 How much money will be required four years from today to repay a $2000 loan that is made today (*a*) at 8% interest, compounded annually? (*b*) at 8% simple interest? *Ans.* (*a*) $2720.97; (*b*) $2640.00

1.24 How many years will be required for an investment of $3000 to increase to $4081.47 at an interest rate of 8% per year, compounded annually? *Ans.* 4 years

1.25 What is the present value of $10 000 to be received 20 years from now, if the principal is invested at 8% per year, compounded annually? *Ans.* $2145.48

1.26 How many years will it take for an investment to double, if the interest rate is 8% per year, compounded annually? *Ans.* 9 years

1.27 A person lends $2000 for five years at 10% per annum simple interest; then the entire proceeds are invested for 10 years at 9% per year, compounded annually. How much money will the person have at the end of the entire 15-year period? *Ans.* $7102.09

1.28 Suppose that a person invests $3000 at 10% per year, compounded annually, for 8 years. (*a*) Will this effectively protect the purchasing power of the original principal, given an annual inflation rate of 8%? (*b*) If so, by how much? *Ans.* (*a*) yes; (*b*) $474.34

1.29 Let the person in Problem 1.28 be in the 45% tax bracket and pay taxes on all the interest received. (*a*) Will the after-tax purchasing power of the original principal be protected? (*b*) Why? *Ans.* (*a*) no; (*b*) $512.57 in purchasing power will be lost

1.30 What amount of money is equivalent to receiving $5000 two years from today, if interest is compounded quarterly at the rate of $2\frac{1}{2}$% per quarter? *Ans.* $4103.73

1.31 On the first day of the year, a man deposits $1000 in a bank at 8% per year, compounded annually. He withdraws $80.00 at the end of the first year, $90.00 at the end of the second year, and the remaining balance at the end of the third year. (*a*) How much does he withdraw at the end of the third year? (*b*) What is his net cash flow? (*c*) How much better off, in terms of net cash flow, would he have been if he had not made the withdrawals at the ends of years one and two? *Ans.* (*a*) $1069.19; (*b*) $239.19; (*c*) $20.51

Chapter 2

Annual Compounding

2.1 SINGLE-PAYMENT, COMPOUND-AMOUNT FACTOR

Suppose that a given sum of money, P, earns interest at a rate i, compounded annually. We have already seen (Section 1.4) that the total amount of money, F, which will have accumulated from an investment of P dollars after n years is given by $F = P(1+i)^n$. The ratio

$$F/P = (1+i)^n \tag{2.1}$$

is called the *single-payment, compound-amount factor.* Numerical values of this factor may be calculated from (*2.1*) or obtained from compound interest tables such as those shown in Appendix A.

A fuller notation, $(F/P, i\%, n)$, is helpful when setting up the solution to a compound interest problem.

Example 2.1 A student deposits \$1000 in a savings account that pays interest at the rate of 6% per year, compounded annually. If all of the money is allowed to accumulate, how much money will the student have after 12 years?

We wish to solve for F, given P, i, and n. Thus,

$$F = P \times (F/P, i\%, n) = \$1000(F/P, 6\%, 12) = \$1000(2.0122) = \$2012.20$$

where the factor $(F/P, 6\%, 12)$ was evaluated from Appendix A.

2.2 SINGLE-PAYMENT, PRESENT-WORTH FACTOR

The *single-payment, present-worth factor* is the reciprocal of the single-payment, compound-amount factor:

$$P/F = (F/P)^{-1} = (1+i)^{-n} \tag{2.2}$$

The expanded notation for this quantity is $(P/F, i\%, n)$. Numerical values for the single-payment, present-worth factor can be obtained directly from (*2.2*) or from a set of tables such as those given in Appendix A.

Example 2.2 A certain sum of money will be deposited in a savings account that pays interest at the rate of 6% per year, compounded annually. If all of the money is allowed to accumulate, how much must be deposited initially so that \$5000 will have accumulated after 10 years?

We wish to solve for P, given F, i, and n. Thus,

$$P = F \times (P/F, i\%, n) = \$5000(P/F, 6\%, 10) = \$5000(0.5584) = \$2792.00$$

where Appendix A gives $(F/P, 6\%, 10)^{-1} = (1.7908)^{-1} = 0.5584$.

2.3 UNIFORM-SERIES, COMPOUND-AMOUNT FACTOR

Let equal amounts of money, A, be deposited in a savings account (or placed in some other interest-bearing investment) at the *end* of each year, as indicated in Fig. 2-1. If the money earns interest at a rate i, compounded annually, how much money will have accumulated after n years?

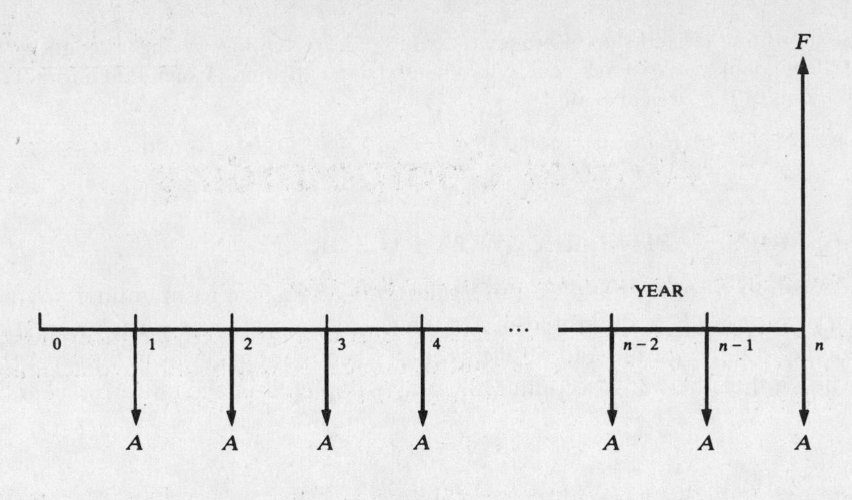

Fig. 2-1

To answer this question, we note that after n years, the first year's deposit will have increased in value to

$$F_1 = A(1+i)^{n-1}$$

Similarly, the second year's deposit will have increased in value to

$$F_2 = A(1+i)^{n-2}$$

and so on. The total amount accumulated will thus be the sum of a geometric progression:

$$\begin{aligned} F &= F_1 + F_2 + \cdots + F_n \\ &= A(1+i)^{n-1} + A(1+i)^{n-2} + \cdots + A \\ &= A[(1+i)^{n-1} + (1+i)^{n-2} + \cdots + 1] \\ &= A\,\frac{(1+i)^n - 1}{i} \end{aligned}$$

The ratio

$$F/A = \frac{(1+i)^n - 1}{i} \tag{2.3}$$

is called the *uniform-series, compound-amount factor.* Numerical values of this factor can be obtained directly, using (*2.3*) in conjunction with an electronic calculator, or from a set of compound interest tables such as those given in Appendix A. The extended notation $(F/A, i\%, n)$ is helpful when solving compound interest problems involving a uniform series.

Example 2.3 A student plans to deposit \$600 each year in a savings account, over a period of 10 years. If the bank pays 6% per year, compounded annually, how much money will have accumulated at the end of the 10-year period?

$$F = A \times (F/A, i\%, n) = \$600(F/A, 6\%, 10) = \$600(13.1808) = \$7908.48$$

2.4 UNIFORM-SERIES, SINKING-FUND FACTOR

The *uniform-series, sinking-fund factor* is the reciprocal of the uniform-series, compound-amount factor:

$$A/F = (F/A)^{-1} = \frac{i}{(1+i)^n - 1} \tag{2.4}$$

This quantity has the extended notation $(A/F, i\%, n)$.

Example 2.4 Suppose that a fixed sum of money, A, will be deposited in a savings account at the end of each year for 20 years. If the bank pays 6% per year, compounded annually, find A such that a total of \$50 000 will be accumulated at the end of the 20-year period.

$$A = F \times (A/F, i\%, n) = \$50\,000(A/F, 6\%, 20) = \$50\,000(0.02718) = \$1359$$

2.5 UNIFORM-SERIES, CAPITAL-RECOVERY FACTOR

Let us now consider a somewhat different situation involving uniform annual payments. Suppose that a given sum of money, P, is deposited in a savings account where it earns interest at a rate i per year, compounded annually. At the end of each year a fixed amount, A, is withdrawn (Fig. 2-2). How large should A be so that the bank account will just be depleted at the end of n years?

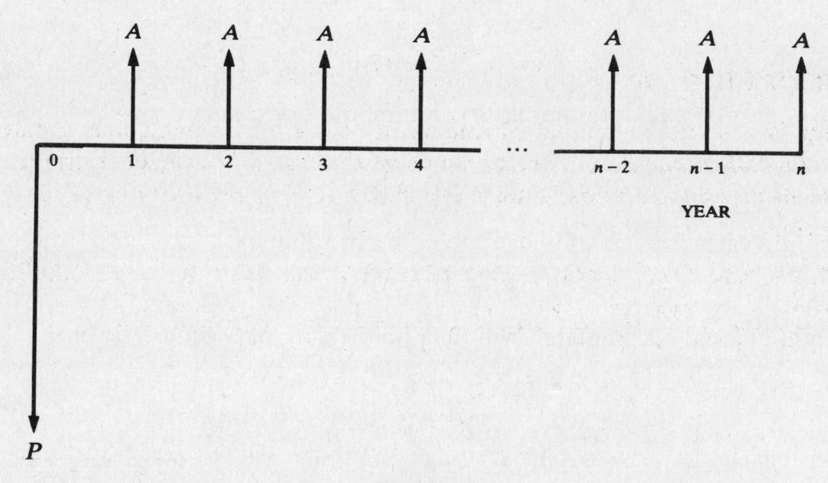

Fig. 2-2

We can make use of previously defined factors to solve this problem, since

$$A \equiv P \times (A/F) \times (F/P) \tag{2.5}$$

Substituting (*2.4*) and (*2.1*) into (*2.5*), we obtain

$$A = P\left[\frac{i}{(1+i)^n - 1}\right](1+i)^n = P\,\frac{i(1+i)^n}{(1+i)^n - 1}$$

The ratio

$$A/P = \frac{i(1+i)^n}{(1+i)^n - 1} = \frac{i}{1-(1+i)^{-n}} \tag{2.6}$$

is called the *uniform-series, capital-recovery factor.* Numerical values of this factor can be computed using (*2.6*) and an electronic calculator, or they can be obtained from a set of compound interest tables such as those given in Appendix A. Symbolically, the uniform-series, capital-recovery factor is written as $(A/P, i\%, n)$.

Example 2.5 An engineer who is about to retire has accumulated \$50 000 in a savings account that pays 6% per year, compounded annually. Suppose that the engineer wishes to withdraw a fixed sum of money at the end of each year for 10 years. What is the maximum amount that can be withdrawn?

$$A = P \times (A/P, i\%, n) = \$50\,000(A/P, 6\%, 10) = \$50\,000(0.1359) = \$6795$$

2.6 UNIFORM-SERIES, PRESENT-WORTH FACTOR

The *uniform-series, present-worth* factor is the reciprocal of the uniform-series, capital-recovery factor:

$$P/A = (A/P)^{-1} = \frac{(1+i)^n - 1}{i(1+i)^n} = \frac{1-(1+i)^{-n}}{i} \tag{2.7}$$

The extended notation is $(P/A, i\%, n)$.

Example 2.6 An engineer who is planning his retirement has decided that he will have to withdraw \$10 000 from his savings account at the end of each year. How much money must the engineer have in the bank at the start of his retirement, if his money earns 6% per year, compounded annually, and he is planning a 12-year retirement (i.e., 12 annual withdrawals)?

$$P = A \times (P/A, i\%, n) = \$10\,000(P/A, 6\%, 12) = \$10\,000(8.3839) = \$83\,839$$

2.7 GRADIENT SERIES FACTOR

A *gradient series* is a series of annual payments in which each payment is greater than the previous one by a constant amount, G. Let us develop the future worth of a gradient series by visualizing it in terms of its component parts, as in Fig. 2-3. Each level constitutes a uniform series, to which (*2.3*) applies; thus,

$$F = A_0(F/A, i\%, n) + G(F/A, i\%, n-1) + G(F/A, i\%, n-2) + \cdots + G(F/A, i\%, 1)$$
$$= A_0\frac{(1+i)^n - 1}{i} + G\frac{(1+i)^{n-1} - 1}{i} + G\frac{(1+i)^{n-2} - 1}{i} + \cdots + G\frac{(1+i) - 1}{i}$$

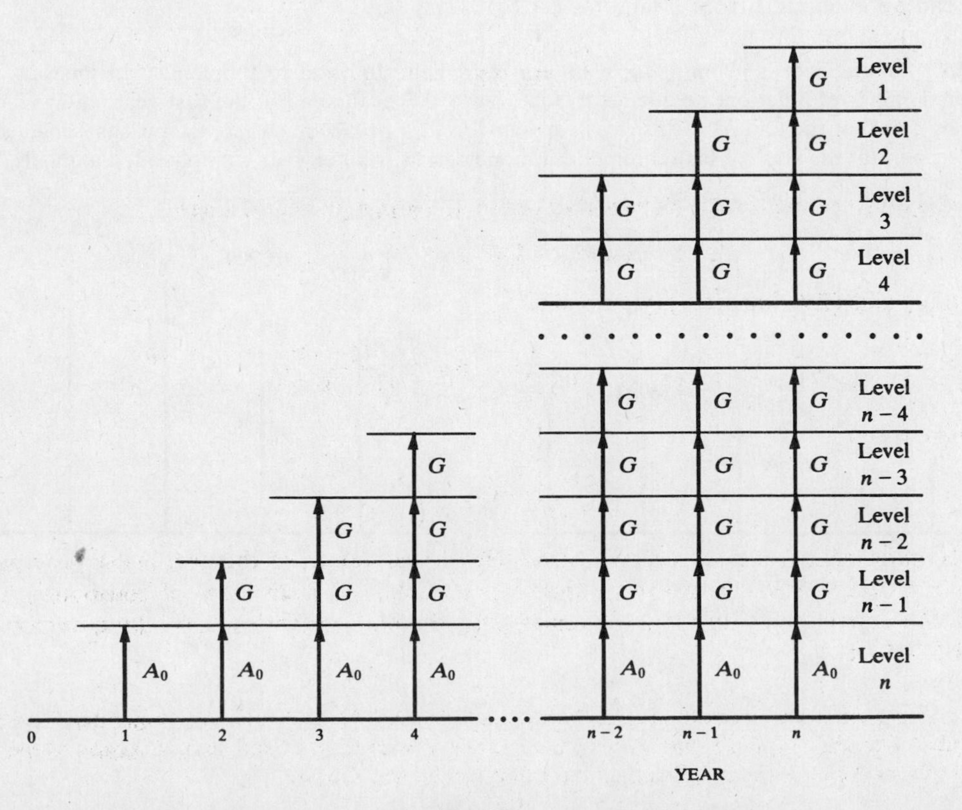

Fig. 2-3

The last $n-1$ terms on the right may be rearranged to give

$$F = A_0 \frac{(1+i)^n - 1}{i} + \frac{G}{i}[(1+i)^{n-1} + (1+i)^{n-2} + \cdots + (1+i) + 1] - \frac{nG}{i}$$

and, summing the geometric progression, we obtain

$$F = A_0 \frac{(1+i)^n - 1}{i} + \frac{G}{i}\left[\frac{(1+i)^n - 1}{i}\right] - \frac{nG}{i} \tag{2.8}$$

Now consider a series of n uniform annual payments, $A_0 + A$, where A_0 has the same value as in the above series of gradients and where A is determined such that the future worth of this uniform series is the same as that given by (*2.8*). Thus,

$$F = (A_0 + A)\left[\frac{(1+i)^n - 1}{i}\right] = \left(A_0 + \frac{G}{i}\right)\left[\frac{(1+i)^n - 1}{i}\right] - \frac{nG}{i}$$

from which

$$A = G\left[\frac{1}{i} - \frac{n}{(1+i)^n - 1}\right]$$

or

$$A/G = \frac{1}{i} - \frac{n}{(1+i)^n - 1} \tag{2.9}$$

Equation (*2.9*) permits direct calculation of the *gradient series factor*, $(A/G, i\%, n)$. Alternatively, since

$$(A/G, i\%, n) = \frac{1}{i} - \frac{n}{i}(A/F, i\%, n) \tag{2.10}$$

the factor can be evaluated from a table of $(A/F, i\%, n)$.

Example 2.7 An engineer is planning for a 15-year retirement. In order to supplement his pension and offset the anticipated effects of inflation, he intends to withdraw \$5000 at the end of the first year, and to increase the withdrawal by \$1000 at the end of each successive year (Fig. 2-4). How much money must the engineer have in his savings account at the start of his retirement, if money earns 6% per year, compounded annually?

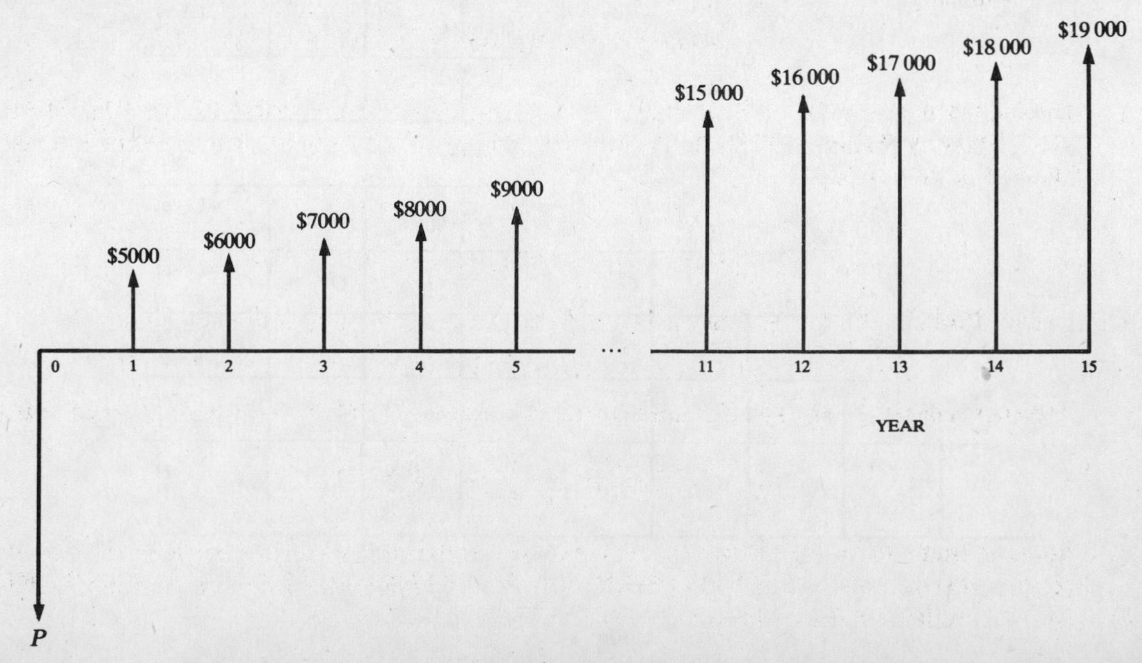

Fig. 2-4

We want to obtain the value of P, given the values of A_0, G, i, and n. We first obtain a series of uniform withdrawals A' equivalent to the series of gradients.

$$A = G \times (A/G, i\%, n) = \$1000(A/G, 6\%, 15) = \$1000(5.9260) = \$5926$$

$$A' = A_0 + A = \$5000 + \$5926 = \$10\,926$$

We can now calculate P as follows:

$$P = A' \times (P/A', i\%, n) = \$10\,926(P/A, 6\%, 15) = \$10\,926(9.7123) = \$106\,116.59$$

A more concise, and the preferable, way to solve this problem is to write

$$\begin{aligned} P &= A_0 \times (P/A, 6\%, 15) + G \times (A/G, 6\%, 15) \times (P/A, 6\%, 15) \\ &= \$5000(9.7123) + \$1000(5.9260)(9.7123) = \$106\,116.59 \end{aligned} \tag{2.11}$$

The gradient series factor can also be used with a *decreasing* series of gradients.

Example 2.8 How much money must initially be deposited in a savings account paying 5% per year, compounded annually, to provide for ten annual withdrawals that start at \$6000 and decrease by \$500 each year?

In this case, $A_0 = \$6000$ and $G = -\$500$. Thus,

$$\begin{aligned} P &= \$6000(P/A, 5\%, 10) - \$500(A/G, 5\%, 10)\,(P/A, 5\%, 10) \\ &= \$6000(7.7217) - \$500(4.0991)\,(7.7217) = \$30\,504.19 \end{aligned}$$

Solved Problems

2.1 A woman deposits \$2000 in a savings account that pays interest at 8% per year, compounded annually. If all the money is allowed to accumulate, how much will she have at the end of (*a*) 10 years? (*b*) 15 years?

(*a*) $$F = P \times (F/P, 8\%, 10) = \$2000(2.1589) = \$4317.80$$

(*b*) $$F = P \times (F/P, 8\%, 15) = \$2000(3.1722) = \$6344.40$$

Alternatively,

$$F = \$4317.80(F/P, 8\%, 5) = \$4317.80(1.4693) = \$6344.14$$

2.2 How much money must be deposited in a savings account so that \$5500 can be withdrawn 12 years hence, if the interest rate is 9% per year, compounded annually, and if all the interest is allowed to accumulate?

$$\begin{aligned} P &= F \times (P/F, 9\%, 12) = F \times (F/P, 9\%, 12)^{-1} \\ &= \$5500(F/P, 9\%, 12)^{-1} = \$5500(2.8127)^{-1} = \$1955.42 \end{aligned}$$

2.3 Repeat Problem 2.2 for an interest rate of $7\frac{1}{2}\%$ per year, compounded annually.

$$P = F \times (P/F, 7\tfrac{1}{2}\%, 12)$$

Appendix A does not include tabular entries for $i = 7\frac{1}{2}\%$ per year; therefore we will make direct use of (*2.2*).

$$P = \frac{\$5500}{(1+0.075)^{12}} = \frac{\$5500}{2.3818} = \$2309.20$$

2.4 Suppose that a person deposits \$500 in a savings account at the end of each year, starting now, for the next 12 years. If the bank pays 8% per year, compounded annually, how much money will accumulate by the end of the 12-year period?

$$F = \$500 \times (F/A, 8\%, 12) = \$500(18.9771) = \$9488.55$$

2.5 Repeat Problem 2.4 for an interest rate of $6\frac{1}{4}\%$ per year, compounded annually.

$$F = \$500(F/A, 6\tfrac{1}{4}\%, 12) = \$500\left[\frac{(1+0.0625)^{12}-1}{0.0625}\right] = \$500(17.1182) = \$8559.12$$

2.6 How much money must be deposited at the end of each year in a savings account that pays 9% per year, compounded annually, in order to have a total of \$10 000 at the end of 14 years?

$$A = F \times (A/F, 9\%, 14) = \$10\,000(F/A, 9\%, 14)^{-1} = \$10\,000(26.0192)^{-1} = \$384.33$$

2.7 A man has deposited \$50 000 in a retirement income plan with a local bank. This bank pays 9% per year, compounded annually, on such deposits. What is the maximum amount the man can withdraw at the end of each year and still have the funds last for 12 years?

From Appendix A,

$$A = \$50\,000(A/P, 9\%, 12) = \$50\,000(0.13965) = \$6982.50$$

2.8 Repeat Problem 2.7 for an interest rate of $8\frac{3}{4}\%$ per year, compounded annually.

This problem must be solved by direct use of *(2.6)*.

$$\begin{aligned} A &= \$50\,000(A/P, 8\tfrac{3}{4}\%, 12) \\ &= \$50\,000\left[\frac{0.0875(1+0.0875)^{12}}{(1+0.0875)^{12}-1}\right] = \$50\,000(0.1379) = \$6894.84 \end{aligned}$$

2.9 Mr. Smith is planning his retirement. He has decided that he needs to withdraw \$12 000 per year from his bank account to supplement his other income from Social Security and a private pension plan. How much money should he plan to have in the bank at the start of his retirement, if the bank pays 10% per year, compounded annually, and if he wants money to last for a 12-year retirement period?

$$P = A \times (P/A, 10\%, 12) = A \times (A/P, 10\%, 12)^{-1} = \$12\,000(0.14676)^{-1} = \$81\,766.15$$

2.10 Mr. Doe is trying to decide whether to put his money in the XYZ Bank or the ABC Bank. The XYZ Bank pays 6% per annum interest, compounded annually; the ABC Bank pays 5% per annum interest, compounded quarterly. Mr. Doe expects to keep his money in the bank for 5 years. Which bank should he select?

From Appendix A,

for XYZ: $(F/P, 6\%, 5) = 1.3382$

for ABC: $(F/P, 1\tfrac{1}{4}\%, 20) = 1.2820$

He should choose XYZ, which offers the greater return per dollar.

2.11 If, in Problem 2.10, Mr. Doe plans to keep his money in the bank for 10 years, is XYZ still the best choice?

for XYZ: $(F/P, 6\%, 10) = 1.7908$

for ABC: $(F/P, 1\tfrac{1}{4}\%, 40) = 1.6436$

It is; in fact, the advantage of XYZ increases with the longer time period.

2.12 Mr. Franklin wants to save for a new sports car that he expects will cost \$38 000 four and one-half years from now. How much money will he have to save each year and deposit in a savings account that pays $6\frac{1}{4}\%$ per year, compounded annually, to buy the car in four and one-half years?

Since the interest is compounded only once a year, Mr. Franklin will have to accumulate the entire \$38 000 during the first four years. Therefore,

$$A = \$38\,000(A/F, 6\tfrac{1}{4}\%, 4) = \$38\,000\left[\frac{0.0625}{(1+0.0625)^4 - 1}\right]$$
$$= \$38\,000(0.2277) = \$8654.32 \text{ per year for four years}$$

2.13 In Problem 2.12, suppose that Mr. Franklin makes a deposit at the *beginning* of each year, rather than at the end. How much money must be deposited each year?

There will now be five deposits, each of size A. At the end of 4 years, the first deposit will have accumulated to

$$A \times (F/P, 6\tfrac{1}{4}\%, 4)$$

and the next four deposits to

$$A \times (F/A, 6\tfrac{1}{4}\%, 4)$$

Hence

$$\$38\,000 = A[(F/P, 6\tfrac{1}{4}\%, 4) + (F/A, 6\tfrac{1}{4}\%, 4)]$$
$$= A(1.2745 + 4.3909) = 5.6654A$$

Solving, $A = \$6707.38$.

2.14 A father wants to set aside money for his 5-year-old son's future college education. Money can be deposited in a bank account that pays 8% per year, compounded annually. What equal deposits should be made by the father, on his son's 6th through 17th birthdays, in order to provide \$5000 on the son's 18th, 19th, 20th, and 21st birthdays?

On the son's 17th birthday, the deposits must have accumulated to

$$P = \$5000(P/A, 8\%, 4)$$
$$= \$5000(A/P, 8\%, 4)^{-1} = \$5000(0.30192)^{-1} = \$16\,560.68$$

Thus, the deposit size, A, must satisfy

$$\$16\,560.68 = A\,(F/A, 8\%, 12)$$
$$\$16\,560.68 = A\,(18.9771)$$
$$A = \$872.67$$

2.15 Dr. Anderson plans to make a series of gradient-type withdrawals from her savings account over a 10-year period, beginning at the end of the second year. What equal annual withdrawals would be equivalent to a withdrawal of \$1000 at the end of the second year, \$2000 at the end of the third year, . . . , \$9000 at the end of the 10th year, if the bank pays 9% per year, compounded annually?

In the notation of Section 2.7, $A_0 = 0$, $G = \$1000$. Hence, from Appendix A,

$$A = \$1000(A/G, 9\%, 10) = \$1000(3.7978) = \$3797.80$$

2.16 Mr. Jones is planning a 20-year retirement; he wants to withdraw \$6000 at the end of the first year, and then to increase the withdrawals by \$800 each year to offset inflation. How much money should he have in his savings account at the start of his retirement, if the bank pays 9% per year, compounded annually, on his savings?

Using an analog of (*2.11*) and the tables in Appendix A,

$$P = A_0\,(P/A, 9\%, 20) + G\,(A/G, 9\%, 20)\,(P/A, 9\%, 20)$$
$$= A_0\,(A/P, 9\%, 20)^{-1} + G\,(A/G, 9\%, 20)\,(A/P, 9\%, 20)^{-1}$$
$$= \$6000(0.10955)^{-1} + \$800(6.7674)\,(0.10955)^{-1} = \$104\,189.14$$

2.17 The ABD Company is building a new plant, whose equipment maintenance costs are expected to be \$500 the first year, \$150 the second year, \$200 the third year, \$250 the fourth year, etc., increasing by \$50 per year through the 10th year. The plant is expected to have a 10-year life. Assuming the interest rate is 8%, compounded annually, how much should the company plan to set aside now in order to pay for the maintenance?

The cash flow diagram is given in Fig. 2-5. First we compute the present worth at the end of year 1 [see (*2.11*)]:

$$\begin{aligned} P' &= \$500 + \$150(P/A, 8\%, 9) + \$50(A/G, 8\%, 9)\,(P/A, 8\%, 9) \\ &= \$500 + [\$150 + \$50(A/G, 8\%, 9)]\,(A/P, 8\%, 9)^{-1} \\ &= \$500 + [\$150 + \$50(3.4910)]\,(0.16008)^{-1} = \$2527.42 \end{aligned}$$

The present worth at the end of year 0 is thus

$$P = P'\,(P/F, 8\%, 1) = \$2527.42(1.0800)^{-1} = \$2340.20$$

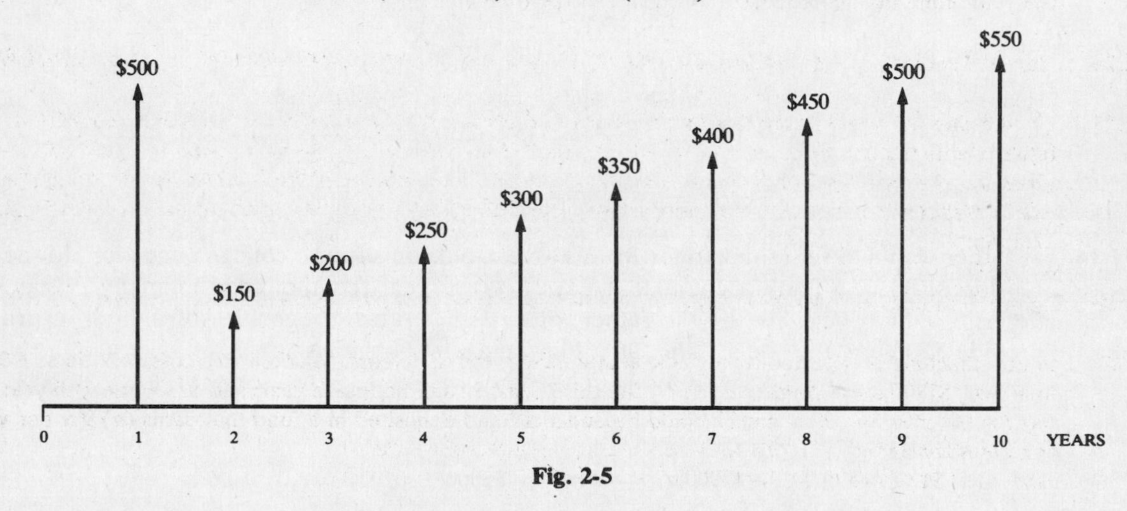

Fig. 2-5

2.18 Slick Oil Company is considering the purchase of a new machine that will last 5 years and cost \$50 000; maintenance will cost \$6000 the first year, decreasing by \$1000 each year to \$2000 the fifth year. If the interest rate is 8% per year, compounded annually, how much money should the company set aside for this machine?

Proceeding as in Problem 2.16, and including the purchase price,

$$\begin{aligned} P &= \$50\,000 + \$6000(A/P, 8\%, 5)^{-1} - \$1000(A/G, 8\%, 5)\,(A/P, 8\%, 5)^{-1} \\ &= \$50\,000 + [\$6000 - \$1000(1.8465)]\,(0.25046)^{-1} = \$66\,583.49 \end{aligned}$$

2.19 Mr. Holzman estimates that the maintenance cost of a new car will be \$75 the first year, and will increase by \$50 each subsequent year. He plans to keep the car for 6 years. He wants to know how much money to deposit in a bank account at the time he purchases the car, in order to cover these maintenance costs. His bank pays $5\frac{1}{2}\%$ per year, compounded annually, on savings deposits.

$$\begin{aligned} P &= \$75(P/A, 5\tfrac{1}{2}\%, 6) + \$50(P/G, 5\tfrac{1}{2}\%, 6) \\ &= [\$75 + \$50(A/G, 5\tfrac{1}{2}\%, 6)]\,(P/A, 5\tfrac{1}{2}\%, 6) \end{aligned}$$

The P/A and A/G factors must be evaluated directly from (*2.7*) and (*2.9*). Thus,

$$(P/A, 5\tfrac{1}{2}\%, 6) = \frac{(1+0.055)^6 - 1}{0.055(1+0.055)^6} = 4.9955$$

$$(A/G, 5\tfrac{1}{2}\%, 6) = \frac{1}{0.055} - \frac{6}{(1+0.055)^6 - 1} = 2.3441$$

and

$$P = [\$75 + \$50(2.3441)]\,(4.9955) = \$960.17$$

Supplementary Problems

2.20 An investment plan pays 15% per year, compounded annually. How much would have to be invested every year so that \$40 000 will be accumulated by the end of 10 years? *Ans.* \$1970.08

2.21 Repeat Problem 2.20 for an interest rate of $13\frac{1}{2}$% per year, compounded annually. *Ans.* \$2119.48

2.22 Mr. Doe borrowed \$1000 from his bank at 8% per year, compounded annually. He can (i) repay the \$1000 together with the interest at the end of 3 years, or (ii) pay the interest at the end of each year and repay the \$1000 at the end of 3 years. By how much is (ii) better than (i)?
Ans. \$259.71 − \$240.00 = \$19.71

2.23 Suppose that \$2000 is invested now, \$2500 two years from now, and \$1200 four years from now, all at 8% per year, compounded annually. What will be the total amount 10 years from now? *Ans.* \$10 849.42

2.24 Repeat Problem 2.23 for an interest rate of $7\frac{3}{4}$% per year, compounded annually. *Ans.* \$10 639.21

2.25 On the day his son was born, a father decided to establish a fund for his son's college education. The father wants the son to be able to withdraw \$4000 from the fund on his 18th birthday, again on his 19th birthday, again on his 20th birthday, and again on his 21st birthday. If the fund earns interest at 9% per year, compounded annually, how much should the father deposit at the end of each year, up through the 17th year? *Ans.* \$350.49

2.26 Repeat Problem 2.25 for an interest rate of $8\frac{1}{2}$% per year, compounded annually. *Ans.* \$370.95

2.27 A new machine is expected to cost \$6000 and have a life of 5 years. Maintenance costs will be \$1500 the first year, \$1700 the second year, \$1900 the third year, \$2100 the fourth year, and \$2300 the fifth year. To pay for the machine, how much should be budgeted and deposited in a fund that earns (*a*) 9% per year, compounded annually? (*b*) $10\frac{1}{2}$% per year, compounded annually?
Ans. (*a*) \$13 256.69; (*b*) \$12 962.59

2.28 The ABC Company has contracted to make the following payments: \$10 000 immediately; \$1000 at the end of year 1; \$1500 at the end of year 2; \$2000 at the end of year 3; \$2500 at the end of year 4; \$3000 at the end of year 5. What fixed amount of money should the company plan to set aside each year, at 8% interest per year, compounded annually, in order to make the above payments? *Ans.* \$4427.82

2.29 Suppose that someone deposits \$2500 in a savings account at the end of each year for the next 15 years. How much money will the person have by the end of the 15th year if the bank pays (*a*) 8%, (*b*) $6\frac{3}{4}$%, per year, compounded annually? *Ans.* (*a*) \$67 880.28; (*b*) \$61 626.00

2.30 Mr. Jones has deposited his life savings of \$70 000 in a retirement income plan with a local bank. The bank pays (*a*) 10%, (*b*) $11\frac{1}{4}$%, per year, compounded annually, on such deposits. What is the maximum fixed amount Mr. Jones can withdraw at the end of each year and still have the funds last for 15 years?
Ans. (*a*) \$9203.16; (*b*) \$9869.27

2.31 Mr. White is planning to take early retirement. He has decided that he needs \$15 000 per year to live on, for the first 5 years of his retirement; after that, his Social Security and other pension plans will provide him an adequate retirement income. How much money must he have in the bank for his 5-year early retirement period, if the bank pays (*a*) 10%, (*b*) $9\frac{1}{4}$%, per year, compounded annually, on the funds? *Ans.* (*a*) \$56 861.80; (*b*) \$57 968.26

2.32 Ms. Frank is planning for a 25-year retirement period and wishes to withdraw a portion of her savings at the end of each year. She plans to withdraw \$10 000 at the end of the first year, and then to increase the amount of the withdrawal by \$1000 each year, to offset inflation. How much money should she have in her savings account at the start of the retirement period, if the bank pays (*a*) 9%, (*b*) $7\frac{1}{2}$%, per year, compounded annually? *Ans.* (*a*) \$175 152.28; (*b*) \$205 435.72

2.33 How much money would have to be saved at (*a*) 8%, (*b*) $8\frac{1}{4}$%, per year, compounded annually, each year for the next 10 years if $50 000 is needed at the end of the 10th year?
Ans. (*a*) $3451.47; (*b*) $3410.71

2.34 A freshman college student, who owns a car, plans to buy a motorcycle. The student expects to save an increasing amount of money on travel every year he is in college, as he will make less use of his car each year. How much money should he plan to save and put in the bank from his job this summer, in order to pay his travel costs for his remaining 3 years of college? Assume that the bank pays 8% per year, compounded annually, and that his travel costs will be $900 the first year, $700 the second year, and $500 the third year. *Ans.* $1830.39

2.35 A father wants to set aside money for his 8-year-old daughter's future education, by making *monthly* deposits to a bank account that pays 8% per year, compounded *annually*. What equal monthly deposits must the father make—the first 1 month after her 9th birthday and the last on her 17th birthday—in order for her to withdraw $4000 on each of her next four birthdays (the 18th through the 21st)?
Ans. $103.80

2.36 Suppose $5000 is deposited in a savings account that pays interest at 8% per year, compounded annually. If no withdrawals are made, how long will it take to accumulate $12 000?
Ans. 12 years (actually, the amount accumulated at the end of 12 years will be $12 590.85)

2.37 Repeat Problem 2.36 for an interest rate of $6\frac{1}{2}$% per year, compounded annually. *Ans.* 14 years

2.38 Suppose that $1000 is deposited in the bank at the end of each year. How long will it take to accumulate $20 000 if the interest rate is 6% per year, compounded annually?
Ans. 14 years (actually, $21 015.07 will have accumulated at the end of 14 years)

2.39 Repeat Problem 2.38 for an interest rate of $5\frac{1}{4}$% per year, compounded annually. *Ans.* 15 years

2.40 Ms. Brown deposits $750 in a savings account at the *beginning* of each year, starting now, for the next 10 years. If the bank pays (*a*) 7%, (*b*) $5\frac{3}{4}$%, per year, compounded annually, how much money will Ms. Brown have accumulated by the end of the 10th year? *Ans.* (*a*) $11 087.70; (*b*) $10 332.09

Chapter 3

Algebraic Relationships and Solution Procedures

3.1 RELATIONSHIPS BETWEEN INTEREST FACTORS

The following relationships are sometimes helpful in interest calculations, particularly when interest tables are used.

$$(F/P, i\%, n) = (F/P, i\%, n_1) \times (F/P, i\%, n_2) \qquad (n = n_1 + n_2) \tag{3.1}$$

$$(P/F, i\%, n) = (P/F, i\%, n_1) \times (P/F, i\%, n_2) \qquad (n = n_1 + n_2) \tag{3.2}$$

$$(F/A, i\%, n) = (F/A, i\%, n_1) + (F/P, i\%, n_1) + (F/P, i\%, n_1 + 1) + \cdots + (F/P, i\%, n - 1) \qquad (n > n_1) \tag{3.3}$$

$$(P/A, i\%, n) = (P/A, i\%, n_1) + (P/F, i\%, n_1 + 1) + (P/F, i\%, n_1 + 2) + \cdots + (P/F, i\%, n) \qquad (n > n_1) \tag{3.4}$$

$$(A/P, i\%, n) = (A/F, i\%, n) + i \tag{3.5}$$

$$\lim_{n \to \infty} (A/P, i\%, n) = i \tag{3.6}$$

Example 3.1 Determine the value of $(F/P, 8\%, 37)$, using the tables presented in Appendix A.

The numerical value of $(F/P, 8\%, 37)$ is not included in the tables; however, from (*3.1*),

$$(F/P, 8\%, 37) = (F/P, 8\%, 30) \times (F/P, 8\%, 7) = 10.0627 \times 1.7138 = 17.2455$$

Example 3.2 Determine the value of $(F/A, 8\%, 37)$, using the tables presented in Appendix A.

The numerical value of $(F/A, 8\%, 37)$ is not included in the tables; however, from (*3.3*) and (*3.1*),

$$\begin{aligned}(F/A, 8\%, 37) &= (F/A, 8\%, 35) + (F/P, 8\%, 35) + (F/P, 8\%, 36)\\ &= (F/A, 8\%, 35) + (F/P, 8\%, 35) + (F/P, 8\%, 35)(F/P, 8\%, 1)\end{aligned}$$

All of the right-hand terms can now be evaluated using the tables in Appendix A. Thus,

$$(F/A, 8\%, 37) = 172.3168 + 14.7853 + (14.7853)(1.08) = 203.0702$$

Example 3.3 Estimate the value of $(A/P, 5\frac{1}{2}\%, 63)$.

From (*3.6*), $(A/P, 5\frac{1}{2}\%, 63) \approx 0.055$. (The correct value, to four decimal places, is 0.0570.)

In addition to the above relationships, there are a number of others that have been discussed in Chapter 2. Specifically, the reader is reminded of the *reciprocal relationships*

$$(P/F, i\%, n) = (F/P, i\%, n)^{-1}$$
$$(A/F, i\%, n) = (F/A, i\%, n)^{-1}$$
$$(A/P, i\%, n) = (P/A, i\%, n)^{-1}$$

and various *product relationships* such as

$$(A/P, i\%, n) = (A/F, i\%, n) \times (F/P, i\%, n)$$
$$(F/G, i\%, n) = (F/A, i\%, n) \times (A/G, i\%, n)$$

3.2 LINEAR INTERPOLATION

If a required compound interest factor falls between two tabulated values, it may be desirable or necessary to use linear interpolation to approximate that factor. Let (x_1, y_1) and (x_2, y_2) be known tabulated points. We wish to determine the value of y corresponding to some given value x, where $x_1 < x < x_2$. Then, by direct proportionality, we can write

$$\frac{y - y_1}{x - x_1} \approx \frac{y_2 - y_1}{x_2 - x_1}$$

Solving for y,

$$y \approx y_1 + \frac{y_2 - y_1}{x_2 - x_1}(x - x_1) \tag{3.7}$$

or, as more commonly applied in practice,

$$y \approx y_1 + \frac{x - x_1}{x_2 - x_1}(y_2 - y_1) \tag{3.8}$$

Example 3.4 Approximate $(F/P, 8\%, 37)$, using the tabulated values in Appendix A and linear interpolation.

From Appendix A, $(F/P, 8\%, 35) = 14.7853$ and $(F/P, 8\%, 40) = 21.7245$. Therefore,

$$\begin{aligned}(F/P, 8\%, 37) &\approx (F/P, 8\%, 35) + \frac{37-35}{40-35}[(F/P, 8\%, 40) - (F/P, 8\%, 35)] \\ &= 14.7853 + \frac{2}{5}(21.7245 - 14.7853) = 17.5610\end{aligned}$$

The correct answer, from (*2.1*), is 17.2456.

Example 3.5 Approximate $(A/P, 8\%, 37)$, using the tabulated values in Appendix A and linear interpolation.

From Appendix A, $(A/P, 8\%, 35) = 0.08580$ and $(A/P, 8\%, 40) = 0.08386$. Therefore,

$$\begin{aligned}(A/P, 8\%, 37) &\approx 0.08580 + \frac{37-35}{40-35}(0.08386 - 0.08580) \\ &= 0.08580 + (0.4)(-0.0019) = 0.08500\end{aligned}$$

The correct answer, from (*2.6*), is 0.08492.

Example 3.6 Approximate $(F/A, 6\frac{3}{4}\%, 15)$ by interpolating linearly between the tabulated values given in Appendix A.

From Appendix A, $(F/A, 6\%, 15) = 23.2760$ and $(F/A, 7\%, 15) = 25.1290$ (note that these values come from two different tables). Hence,

$$\begin{aligned}(F/A, 6\tfrac{3}{4}\%, 15) &\approx (F/A, 6\%, 15) + \frac{6.75-6.00}{7.00-6.00}[(F/A, 7\%, 15) - (F/A, 6\%, 15)] \\ &= 23.2760 + (0.75)(25.1290 - 23.2760) = 24.6658\end{aligned}$$

The correct answer, from (*2.3*), is 24.6504.

3.3 UNKNOWN NUMBER OF YEARS

Sometimes we require the number of years, n, that corresponds to a given compound interest factor and a given annual interest rate, i. Solving the appropriate formulas from Chapter 2 by logarithms, we obtain the results displayed in Table 3-1. Formula (*2.9*) cannot be solved exactly for n, given A/G and i; in this case, a graphical solution or linear interpolation is employed.

Table 3-1

Factor	Number of Years
F/P or P/F	$n = \dfrac{\log (F/P)}{\log (1+i)} = \dfrac{-\log (P/F)}{\log (1+i)}$
F/A or A/F	$n = \dfrac{\log [1+i(F/A)]}{\log (1+i)} = \dfrac{\log \left(1+\dfrac{i}{A/F}\right)}{\log (1+i)}$
P/A or A/P	$n = \dfrac{-\log [1-i(P/A)]}{\log (1+i)} = \dfrac{-\log \left(1-\dfrac{i}{A/P}\right)}{\log (1+i)}$

Example 3.7 How many years will be required for a given sum of money to triple, if it is deposited in a bank account that pays 6% per year, compounded annually?

We require n, given $F/P = 3$ and $i = 0.06$. From Table 3-1,

$$n = \frac{\log 3}{\log (1+0.06)} = 18.85$$

But, since the interest is compounded only at the end of each year, the calculated value for n must be an integer. Hence $n = 19$ (corresponding to $F/P = 3.0256$) is the correct solution.

Example 3.8 Determine the value of n corresponding to $A/F = 0.01$, if $i = 7\frac{1}{2}\%$ per year, compounded annually.

From Table 3-1,

$$n = \frac{\log \left(1+\dfrac{0.075}{0.01}\right)}{\log (1+0.075)} = \frac{\log 8.5}{\log 1.075} = 29.59$$

The required number of years is therefore 30.

Example 3.9 Determine the value of n corresponding to $A/G = 11.6$ and $i = 7\%$, compounded annually, using the following known points $(n, A/G)$ (all at $i = 7\%$): (35, 10.6687), (40, 11.4234), (45, 12.0360).

From the graph of the data, Fig. 3-1, we can see that the value of n corresponding to $A/G = 11.6$ is approximately 41.4.

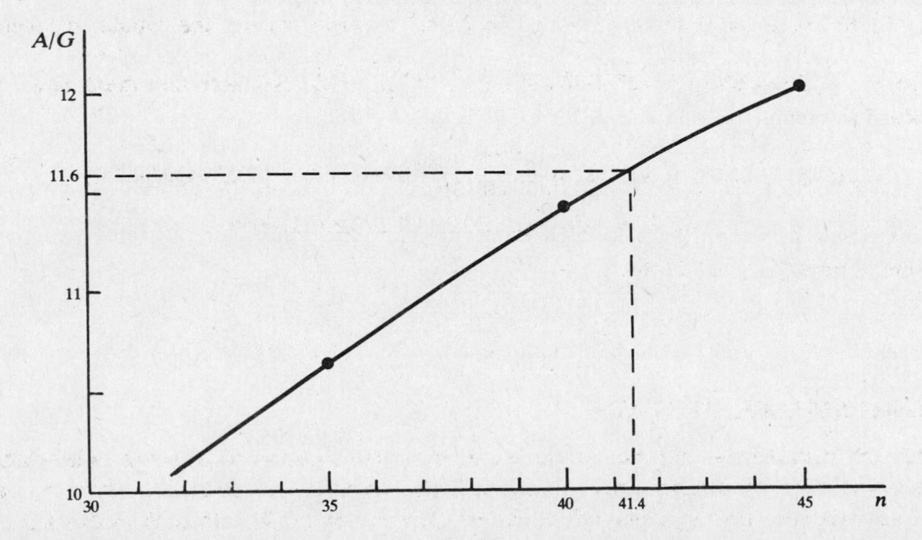

Fig. 3-1

The value of n can also be approximated by interpolating linearly between the second two data points:

$$n \approx 40 + \frac{11.6000 - 11.4234}{12.0360 - 11.4234}(45 - 40) = 41.44$$

Since the interest is compounded annually, n can only take on integer values. Thus, $n = 41$ comes closest to satisfying the given condition that $A/G = 11.6$.

3.4 UNKNOWN INTEREST RATE

Another situation that frequently arises is the need to solve for the annual interest rate that corresponds to a given compound interest factor and a specified number of years. If the given compound interest factor is either F/P or P/F, then the value of i can be obtained explicitly as

$$i = (F/P)^{1/n} - 1 = (P/F)^{-1/n} - 1 \tag{3.9}$$

When one of the remaining five interest factors is specified, however, it is necessary to utilize a numerical or graphical solution procedure, or to interpolate linearly between tabulated values. If the given interest factor is not tabulated for the specified value of n, double interpolation will be required.

Example 3.10 Determine the value of i corresponding to $F/A = 1000$ and $n = 38$, using the known values displayed in Table 3-2.

Table 3-2

i	n	F/A
12%	35	431.6635
	40	767.0914
	45	1358.2300
15%	35	881.1702
	40	1779.0904
	45	3585.1286

We first obtain an interpolated value of F/A for $i = 12\%$ and $n = 38$:

$$(F/A, 12\%, 38) \approx 431.6635 + \frac{38 - 35}{40 - 35}(767.0914 - 431.6635) = 632.9202$$

Then we obtain an interpolated value of F/A for $i = 15\%$ and $n = 38$:

$$(F/A, 15\%, 38) \approx 881.1702 + \frac{38 - 35}{40 - 35}(1779.0904 - 881.1702) = 1419.9223$$

We can now interpolate between these calculated values to obtain the desired interest rate:

$$i \approx 12 + \frac{1000.0000 - 632.9202}{1419.9223 - 632.9202}(15 - 12) = 13.40\%$$

A more accurate result can be obtained if the factors $(F/A, 12\%, 38)$ and $(F/A, 15\%, 38)$ are evaluated directly from (*2.3*). Thus,

$$(F/A, 12\%, 38) = \frac{(1 + 0.12)^{38} - 1}{0.12} = 609.8305$$

$$(F/A, 15\%, 38) = \frac{(1 + 0.15)^{38} - 1}{0.15} = 1343.6222$$

and now linear interpolation gives:

$$i \approx 12 + \frac{1000.0000 - 609.8305}{1343.6222 - 609.8305}(15 - 12) = 13.60\%$$

This latter method is preferable, provided a calculator is available to carry out the exponentiation.

Solved Problems

3.1 Evaluate (*a*) $(F/P, 10\%, 44)$, (*b*) $(A/F, 10\%, 37)$.

(*a*) $$(F/P, 10\%, 44) = (F/P, 10\%, 40) \times (F/P, 10\%, 4) = (45.2593)(1.4641) = 66.2641$$

(*b*) By (*3.3*) and (*3.1*),

$$\begin{aligned}(F/A, 10\%, 37) &= (F/A, 10\%, 35) + (F/P, 10\%, 35) + [(F/P, 10\%, 35) \times (F/P, 10\%, 1)] \\ &= 271.0244 + 28.1024 + (28.1024)(1.1000) = 330.039\end{aligned}$$

Hence, $(A/F, 10\%, 37) = 1/330.039 = 0.00303$.

3.2 Estimate (*a*) $(A/P, 8\%, 100)$, (*b*) $(P/A, 20\%, 50)$.

(*a*) By (*3.6*), $(A/P, 8\%, 100) \approx 0.08$ (Appendix A gives the correct value as 0.08004).

(*b*) $$(P/A, 20\%, 50) = (A/P, 20\%, 50)^{-1} \approx (0.20)^{-1} = 5$$

The correct value is 4.9995.

3.3 Find $(A/F, i\%, n)$ from (*a*) $(F/P, i\%, n)$, (*b*) $(P/F, i\%, n)$, (*c*) $(A/P, i\%, n)$.

(*a*) $$(A/F, i\%, n) = \frac{i}{(F/P, i\%, n) - 1}$$

(*b*) $$(A/F, i\%, n) = \frac{i(P/F, i\%, n)}{1 - (P/F, i\%, n)}$$

(*c*) $$(A/F, i\%, n) = (A/P, i\%, n) - i$$

3.4 Derive (*3.1*) without reference to (*2.1*).

Investing \$1 for n_1 years at $i\%$ annual compound interest, and then investing the accumulated amount for a further n_2 years at the same rate, must be equivalent to investing \$1 for $n_1 + n_2$ years at $i\%$.

3.5 Derive (*3.3*) without reference to (*2.3*) or (*2.1*).

As shown in Fig. 3-2, a uniform series of n payments of \$1 may be considered as composed of two parts: the first $n - n_1$ payments, whose total future worth at the end of year n is

$$(F/P, i\%, n-1) + (F/P, i\%, n-2) + \cdots + (F/P, i\%, n_1) \qquad (1)$$

and the uniform series of the remaining n_1 payments, whose future worth is

$$(F/A, i\%, n_1) \qquad (2)$$

The sum of (*1*) and (*2*) must be the future worth of the entire series, $(F/A, i\%, n)$.

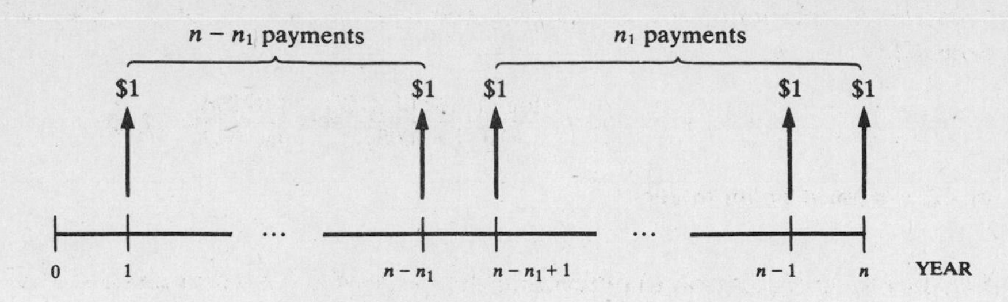

Fig. 3-2

It is obvious from Fig. 3-2 that the sum (*1*) above is equal to

$$(F/A, i\%, n - n_1)\,(F/P, i\%, n_1)$$

Hence we have the following, more compact version of (*3.3*):

$$(F/A, i\%, n) = (F/A, i\%, n_1) + (F/A, i\%, n - n_1)\,(F/P, i\%, n_1) \qquad (n > n_1) \tag{3.10}$$

3.6 Rework Problem 3.1(*b*), using (*3.10*).

$$\begin{aligned}(F/A, 10\%, 37) &= (F/A, 10\%, 35) + (F/A, 10\%, 2)\,(F/P, 10\%, 35)\\ &= 271.0244 + (2.1000)(28.1024) = 330.039\end{aligned}$$

as before.

3.7 Use linear interpolation and the tables given in Appendix A to evaluate (*a*) $(P/F, 6\%, 52)$, (*b*) $(P/A, 8\frac{1}{4}\%, 10)$, (*c*) $(A/G, 13\%, 48)$.

(*a*) First interpolate for $(F/P, 6\%, 52)$, then take the reciprocal.

$$(F/P, 6\%, 52) \approx 18.4202 + \frac{52 - 50}{55 - 50}(24.6503 - 18.4204) = 20.9122$$

$$(P/F, 6\%, 52) \approx (20.9122)^{-1} = 0.04782$$

(*b*) First interpolate for $(A/P, 8\frac{1}{4}\%, 10)$, then take the reciprocal.

$$(A/P, 8\tfrac{1}{4}\%, 10) \approx 0.14903 + \frac{8.25 - 8.00}{9.00 - 8.00}(0.15582 - 0.14903) = 0.15073$$

$$(P/A, 8\tfrac{1}{4}\%, 10) \approx (0.15073)^{-1} = 6.6345$$

(*c*) First interpolate on n to obtain $(A/G, 12\%, 48)$ and $(A/G, 15\%, 48)$; then interpolate on i to obtain $(A/G, 13\%, 48)$.

$$(A/G, 12\%, 48) \approx 8.0572 + \frac{48 - 45}{50 - 45}(8.1597 - 8.0572) = 8.1187$$

$$(A/G, 15\%, 48) \approx 6.5830 + \frac{48 - 45}{50 - 45}(6.6205 - 6.5830) = 6.6055$$

$$(A/G, 13\%, 48) \approx 8.1187 + \frac{13 - 12}{15 - 12}(6.6055 - 8.1187) = 7.6143$$

3.8 How many years will be required for a sum of money to quadruple, if it is deposited in a bank account that pays $6\frac{3}{4}\%$ per year, compounded annually?

$$n = \frac{\log 4}{\log(1 + 0.0675)} = \frac{0.60206}{0.02837} = 21.22 \text{ years}$$

or 22 years, if n must be an integer.

3.9 Determine the value of n corresponding to $P/A = 12$, if $i = 8\%$ per year, compounded annually.

$$n = \frac{-\log[1-(0.08)(12)]}{\log(1+0.08)} = \frac{1.39794}{0.03342} = 41.82$$

or 42, if n must be an integer.

3.10 Use linear interpolation to determine the value of n corresponding to $A/G = 5.4000$ and $i = 8\%$ per year, compounded annually.

From the tables in Appendix A,

$$(A/G, 8\%, 14) = 5.2731 \qquad (A/G, 8\%, 15) = 5.5945$$

Then

$$n \approx 14 + \frac{5.4000 - 5.2731}{5.5945 - 5.2731}(15-14) = 14.39 \text{ years}$$

3.11 A bank will return \$2345 on a 10-year certificate of deposit that originally cost \$1000. What interest rate, compounded annually, is the bank paying?

$$i = \left(\frac{\$2345}{\$1000}\right)^{1/10} - 1 = 1.0899 - 1 = 8.89\%$$

3.12 By interpolation, determine the value of i corresponding to $A/P = 0.15000$ and $n = 12$.

From the tables in Appendix A, we find that for $n = 12$:

i	A/P
10%	0.14676
12%	0.16144

Therefore, for $A/P = 0.15000$,

$$i \approx 10\% + \frac{0.15000 - 0.14676}{0.16144 - 0.14676}(12\% - 10\%) = 10.44\%$$

Supplementary Problems

3.13 Evaluate, using (*3.1*) and Appendix A, (*a*) $(F/P, 10\%, 41)$, (*b*) $(F/P, 12\%, 43)$, (*c*) $(F/P, 6\%, 57)$. *Ans.* (*a*) 49.7852; (*b*) 130.7299; (*c*) 27.6971

3.14 Evaluate, using (*3.3*) or (*3.10*) and the tables in Appendix A, (*a*) $(F/A, 10\%, 41)$, (*b*) $(F/A, 12\%, 43)$, (*c*) $(F/A, 6\%, 57)$. *Ans.* (*a*) 487.8518; (*b*) 1081.0826; (*c*) 444.9517

3.15 Evaluate, using (*3.4*) and Appendix A, (*a*) $(P/A, 5\%, 52)$, (*b*) $(P/A, 9\%, 39)$, (*c*) $(P/A, 12\%, 43)$. *Ans.* (*a*) 18.4181; (*b*) 10.7255; (*c*) 8.2696

3.16 Evaluate, using (*3.5*) and the tabular values of $(A/P, i\%, n)$ given in Appendix A, (*a*) $(A/F, 4\%, 20)$, (*b*) $(A/F, 15\%, 8)$, (*c*) $(A/F, 12\%, 40)$. *Ans.* (*a*) 0.03358; (*b*) 0.07285; (*c*) 0.00130

3.17 Estimate the following factors, using (*3.6*), and compare with the correct values as given by (*2.6*): (*a*) $(A/P, 8\%, 80)$, (*b*) $(A/P, 4\frac{3}{4}\%, 120)$. (*c*) $(A/P, 19.08\%, 100)$.
Ans. (*a*) 0.08 (correct value, 0.0802); (*b*) 0.0475 (correct value, 0.0477); (*b*) 0.1908 (correct value, 0.1908)

3.18 From the interest tables in Appendix A, determine the value of each of the following compound interest factors using linear interpolation. Compare each value with the exact answer obtained from the appropriate formula in Chapter 2. (*a*) $(F/P, 7\frac{1}{2}\%, 12)$, (*b*) $(F/P, 12\%, 47)$, (*c*) $(P/F, 5\frac{1}{4}\%, 20)$, (*d*) $(F/A, 4\%, 38)$, (*e*) $(F/A, 4\frac{3}{4}\%, 38)$, (*f*) $(A/F, 11\frac{1}{4}\%, 30)$, (*g*) $(A/P, 5\%, 57)$, (*h*) $(A/P, 7\frac{1}{4}\%, 57)$, (*i*) $(P/A, 10\frac{3}{4}\%, 30)$, (*j*) $(A/G, 8\frac{1}{4}\%, 15)$, (*k*) $(A/G, 13\frac{3}{4}\%, 39)$.

Ans. (*a*) 2.3852 (exact: 2.3818); (*b*) 213.9934 (205.7061); (*c*) 0.3606 (0.3594); (*d*) 86.4762 (85.9703); (*e*) 92.0092 (101.7364); (*f*) 0.004869 (0.00479); (*g*) 0.05333 (0.05330); (*h*) 0.07391 (0.07387); (*i*) 8.9125 (8.8675); (*j*) 5.5545 (5.5541); (*k*) 7.0591 (7.0146)

3.19 How many years (a whole number) will be required for a sum of money to double if the interest rate is (*a*) 10%, (*b*) 12%, (*c*) $13\frac{1}{2}\%$, per year, compounded annually?
Ans. (*a*) 8 years; (*b*) 7 years; (*c*) 6 years

3.20 How many years (a whole number) will be required to accumulate $10 000 if $500 is deposited at the end of each year, and interest is payable at $6\frac{3}{4}\%$ per year, compounded annually? *Ans.* 14 years

3.21 A person has $80 000 in a savings account that earns interest at $7\frac{1}{2}\%$ per year, compounded annually. If the person withdraws $12 000 at the end of each year, after how many years (a whole number) will the savings be exhausted? *Ans.* 10 years

3.22 Use linear interpolation to determine the value of n corresponding to $A/G = 6.0000$, if $i = 9\%$ per year, compounded annually. *Ans.* 16.99

3.23 Repeat Problem 3.22 for an interest rate of 11% per year, compounded annually. *Ans.* 18.84

3.24 A man has entered into a contract in which he has agreed to lend $1000 to a friend, and the friend has agreed to repay him $1060.90 two years later. What annual compound rate of interest is the man receiving on his $1000? *Ans.* 3%

3.25 The First National Bank advertises it will pay $3869.70 in cash at the end of 20 years to anyone who deposits $1000. Federal Savings, a competitor, advertises that it pays 10% per year, compounded annually, on all deposits left one year or more. Which bank is paying the higher interest rate, and by how much? *Ans.* Federal Savings, by 3%

3.26 A person who is about to retire has accumulated $100 000 in a savings account. Suppose that the person withdraws $8195.23 from the savings account at the end of each year for 20 years, at which time the account is totally depleted. What is the interest rate, based upon annual compounding?
Ans. $5\frac{1}{4}\%$ per year

3.27 A person is considering entering into an agreement with an investment company to deposit $1000 into a special account at the end of each year for 15 years. At the end of the period, the person would be able to withdraw a lump sum of $28 800. At what rate would the person earn interest, if the interest was compounded annually? *Ans.* $8\frac{3}{4}\%$ per year

3.28 A person is considering entering into an agreement with an investment company to deposit $1000 into a special account at the end of the first year, $1100 at the end of the second year, etc., increasing by $100 each year. At the end of 15 years the person would be able to withdraw a lump sum of $36 000. At what rate would the person earn interest, if the interest was compounded annually?
Ans. 5.54% per year

Chapter 4

Discrete, Periodic Compounding

4.1 NOMINAL AND EFFECTIVE INTEREST RATES

Many financial transactions require that interest be compounded more often than once a year (e.g., quarterly, monthly, daily, etc.). In such situations, there are two expressions for the interest rate. The *nominal* interest rate, r, is expressed on an annual basis; this is the rate that is normally quoted when describing an interest-bearing transaction. The *effective* interest rate, i, is the rate that corresponds to the actual interest period. The effective interest rate is obtained by dividing the nominal interest rate by m, the number of interest periods per year:

$$i = \frac{r}{m} \tag{4.1}$$

Example 4.1 A bank claims to pay interest to its depositors at the rate of 6% per year, compounded quarterly. What are the nominal and effective interest rates?

The nominal interest rate is $r = 6\%$. Since there are four interest periods per year, the effective interest rate is

$$i = \frac{6\%}{4} = 1.5\% \text{ per quarter}$$

4.2 WHEN INTEREST PERIODS COINCIDE WITH PAYMENT PERIODS

When the interest periods and the payment periods coincide, it is possible to make direct use both of the compound interest formulas developed in Chapter 2 and the compound interest tables presented in Appendix A, provided the interest rate, i, is taken to be the *effective* interest rate *for that interest period.* Moreover, the number of years, n, must be replaced by the total number of interest periods, mn.

Example 4.2 An engineer plans to borrow \$3000 from his company credit union, to be repaid in 24 equal monthly installments. The credit union charges interest at the rate of 1% per month on the unpaid balance. How much money must the engineer repay each month?

This problem can be solved by direct application of (*2.6*), since the interest charges and the uniform payments are both determined on a monthly basis:

$$A = P \times (A/P, 1\%, mn) = \$3000\ (A/P, 1\%, 24)$$

$$= \$3000 \frac{0.01(1+0.01)^{24}}{(1+0.01)^{24} - 1} = \$141.22$$

We conclude that the engineer must repay \$141.22 at the end of every month for 24 months.

Alternatively, Appendix A gives $(A/P, 1\%, 24) = 0.04707$, whence

$$A = \$3000(0.04707) = \$141.21 \text{ per month}$$

If, as in the case of commercial loans, the nominal interest rate is specified, the compound interest formulas of Chapter 2 and/or Appendix A can still be used, with i replaced by r/m, and n by mn.

EXAMPLE 4.3 An engineer wishes to purchase an \$80 000 home by making a down payment of \$20 000 and borrowing the remaining \$60 000, which he will repay on a monthly basis over the next 30 years. If the bank charges interest at the rate of $9\frac{1}{2}\%$ per year, compounded monthly, how much money must the engineer repay each month?

Again applying (*2.6*),

$$A = P \times (A/P, r\%/m, mn)$$
$$= \$60\,000 \frac{(0.095/12)(1+0.095/12)^{(12)(30)}}{(1+0.095/12)^{(12)(30)}-1} = \$504.51$$

It is interesting to note that the total amount of money which will be repaid to the bank is

$$\$504.51 \times 360 = \$181\,623.60$$

or *three times* the amount of the original loan.

4.3 WHEN INTEREST PERIODS ARE SMALLER THAN PAYMENT PERIODS

If the interest periods are smaller than the payment periods, then the interest may be compounded several times between payments. One way to handle problems of this type is to determine the effective interest rate for the given interest period, and then treat each payment separately.

Example 4.4 An engineer deposits \$1000 in a savings account at the end of each year. If the bank pays interest at the rate of 6% per year, compounded quarterly, how much money will have accumulated in the account after 5 years?

The effective interest rate is $i = 6\%/4 = 1.5\%$ per quarter; the first deposit accumulates for 16 quarters; etc.

$$F = \$1000(F/P, 1.5\%, 16) + \$1000(F/P, 1.5\%, 12)$$
$$+ \$1000(F/P, 1.5\%, 8) + \$1000(F/P, 1.5\%, 4) + \$1000(F/P, 1.5\%, 0)$$

The F/P factors can be obtained from either (*2.1*) or Appendix A.

$$F = \$1000(1.2690) + \$1000(1.1956) + \$1000(1.1265) + \$1000(1.0614) + \$1000(1.0000)$$
$$= \$5652.50$$

Another procedure, which is usually more convenient, is to calculate an effective interest rate for the given *payment* period, and then to proceed as though the interest periods and the payment periods coincided. This effective interest rate can be determined as

$$i = \left(1 + \frac{r}{\alpha}\right)^{\alpha} - 1 \tag{4.2}$$

where α represents the number of interest periods per payment period and r is the nominal interest rate for that payment period. If the payment period is one year, then $\alpha = m$, and we obtain the following expression for the *effective annual interest rate*:

$$i = \left(1 + \frac{r}{m}\right)^{m} - 1 \tag{4.3}$$

Example 4.5 Rework Example 4.4 by using an effective annual interest rate.

Here, $r = 6\%$ and $\alpha = m = 4$, so that, by (*4.3*),

$$i = \left(1 + \frac{0.06}{4}\right)^{4} - 1 = 0.06136$$

We can now apply (*2.3*) to obtain

$$F = \$1000(F/A, 6.136\%, 5) = \$1000 \frac{(1+0.06136)^5 - 1}{0.06136} = \$5652.40$$

which agrees with Example 4.4 to within roundoff errors.

Appendix B contains a tabulation of effective annual interest rates corresponding to various nominal interest rates. This table may be used in place of (*4.3*), if desired.

4.4 WHEN INTEREST PERIODS ARE LARGER THAN PAYMENT PERIODS

If the interest periods are larger than the payment periods, some of the payments may not have been deposited for an entire interest period. Such payments do not earn any interest during that interest period. In other words, interest is earned only by those payments that have been deposited or invested for the entire interest period.

Situations of this type can be treated in the following manner:

1. Consider all *deposits* that were made during the interest period to have been made at the *end* of the interest period (and therefore to have earned no interest during that interest period).
2. Consider all *withdrawals* that were made during the interest period to have been made at the *beginning* of the interest period (again earning no interest).
3. Then proceed as though the interest periods and the payment periods coincided.

Example 4.6 A person has \$4000 in a savings account at the beginning of a calendar year; the bank pays interest at 6% per year, compounded quarterly. Table 4-1 shows the transactions carried out during the calendar year; the second column gives the effective dates according to rules 1 and 2 above. To find the balance in the account at the end of the calendar year, we calculate the effective interest rate, 6%/4 = 1.5% per quarter. Then, lumping the amounts at the effective dates and applying (*2.1*), we obtain

$$\begin{aligned} F &= (\$4000-\$175)(F/P, 1.5\%, 4) + (\$1200-\$1800)(F/P, 1.5\%, 3) \\ &\quad + (\$180-\$800)(F/P, 1.5\%, 2) + (\$1600-\$1100)(F/P, 1.5\%, 1) + \$2300 \\ &= \$3825(1.0614) - \$600(1.0457) - \$620(1.0302) + \$500(1.0150) + \$2300 \\ &= \$5601.21 \end{aligned}$$

Table 4-1

Date	Effective Date	Deposit	Withdrawal
Jan. 10	Jan. 1		\$ 175
Feb. 20	Mar. 31	\$1200	
Apr. 12	Apr. 1		1500
May 5	June 30	65	
May 13	June 30	115	
May 24	Apr. 1		50
June 21	Apr. 1		250
Aug. 10	Sept. 30	1600	
Sept. 12	July 1		800
Nov. 27	Oct. 1		350
Dec. 17	Dec. 31	2300	
Dec. 29	Oct. 1		750

Solved Problems

4.1 A bank advertises that it pays interest at the rate of 10% per year, compounded quarterly. What effective interest rate is the bank paying?

$$r = 10\% \qquad i = \frac{10\%}{4} = 2.5\% \text{ per quarter}$$

4.2 An engineer has just borrowed $8000 from a local bank, at the rate of 1% per month on the unpaid balance. His contract states that he must repay the loan in 35 equal monthly installments. How much money must he repay each month?

$$A = \$8000(A/P, 1\%, 35) = \$8000(0.03400) = \$272.00$$

4.3 A bank pays interest at the rate of 6% per year, compounded monthly. If a person deposits $2500 in a savings account at the bank, how much money will accumulate by the end of 2 years?

Equation (*2.1*), with the appropriate substitutions, gives

$$F = \$2500\left(1 + \frac{0.06}{12}\right)^{(12)(2)} = \$2500(1.005)^{24} = \$2817.89$$

Alternatively, using the tables in Appendix A, we have

$$F = \$2500(F/P, 0.5\%, 24) = \$2500(1.1272) = \$2818.00$$

4.4 A man plans to buy a $150 000 house. He wants to make a down payment of $30 000 and to take out a 30-year mortgage for the remaining $120 000, at 10% per year, compounded monthly. How much must he repay each month?

Equation (*2.6*), with the appropriate substitutions, gives

$$A = \$120\,000\,\frac{(0.10/12)(1 + 0.10/12)^{(12)(30)}}{(1 + 0.10/12)^{(12)(30)} - 1} = \$120\,000\,\frac{0.16531}{18.8374} = \$1053.08$$

The solution can also be approximated by use of (*3.6*):

$$A = \$120\,000(A/P, 10\%/12, 360) \approx \$120\,000(0.10/12) = \$1000$$

4.5 A man plans to save $1000 a month for the next 20 years, at 10% per year, compounded monthly. How much money will he have at the end of 20 years?

Equation (*2.3*), with the appropriate substitutions, gives

$$F = \$1000\,\frac{(1 + 0.10/12)^{(12)(20)} - 1}{0.10/12} = \$1000\,\frac{6.32807}{0.0083333} = \$759\,371.43$$

4.6 Repeat Problem 4.5 using quarterly compounding.

$$F = \$1000\,\frac{(1 + 0.10/4)^{(4)(20)} - 1}{0.10/4} = \$1000\,\frac{6.20957}{0.025} = \$248\,382.80$$

Note that this is much less than the value obtained using monthly compounding.

4.7 What is the present value of a stream of monthly payments of $500 each over 10 years, if the interest rate is 10% per annum, compounded monthly?

Equation (2.7), with the appropriate substitutions, gives

$$P = \$500\,\frac{(1+0.10/12)^{(12)(10)}-1}{(0.10/12)(1+0.10/12)^{(12)(10)}} = \$500\,\frac{1.70704}{(0.0083333)(2.70704)} = \$37\,835.72$$

4.8 Repeat Problem 4.7 using daily compounding. For computational simplicity, assume 30 days in each month (many banks do this).

Here, $r = 10\%/12 = 0.00833333$ and $\alpha = 30$; hence by (4.2), the effective monthly interest rate is

$$i = \left(1+\frac{0.00833333}{30}\right)^{30} - 1 = 0.0083670$$

Now use (2.7), as before.

$$P = \$500\,\frac{(1+0.0083670)^{(12)(10)}-1}{(0.0083670)(1+0.0083670)^{(12)(10)}} = \$500\,\frac{1.717909}{(0.0083670)(2.717909)} = \$37\,771.61$$

4.9 How much money must be deposited in a savings account each month to accumulate \$10 000 at the end of 5 years, if the bank pays interest at the rate of 6% per year, compounded (*a*) monthly? (*b*) semiannually? (*c*) quarterly? (*d*) daily?

In each case use (2.4), with the appropriate substitutions.

(*a*)
$$A = \$10\,000\,\frac{0.06/12}{(1+0.06/12)^{(12)(5)}-1} = \$10\,000\,\frac{0.005}{0.34885} = \$143.33 \text{ per month}$$

(*b*)
$$A = \$10\,000\,\frac{0.06/2}{(1+0.06/2)^{(2)(5)}-1} = \$10\,000\,\frac{0.03}{0.34392} = \$872.30 \text{ every 6 months}$$

or \$872.30/6 = \$145.38 per month.

(*c*)
$$A = \$10\,000\,\frac{0.06/4}{(1+0.06/4)^{(4)(5)}-1} = \$10\,000\,\frac{0.015}{0.34686} = \$432.45 \text{ per quarter}$$

or \$432.45/3 = \$144.15 per month.

(*d*)
$$r = \frac{6\%}{12} = 0.005 \qquad \alpha = 30$$

$$i = \left(1+\frac{0.005}{30}\right)^{30} - 1 = 0.0050121$$

$$A = \$10\,000\,\frac{0.0050121}{(1+0.0050121)^{(12)(5)}-1} = \$10\,000\,\frac{0.0050121}{0.34982} = \$143.28 \text{ per month}$$

4.10 Using a suitable version of (2.9), evaluate A/G for $r = 12\%$, compounded monthly, and $n = 2$ years. Compare the result with the tabulated value in Appendix A.

$$A/G = \frac{m}{r} - \frac{mn}{(1+r/m)^{mn}-1}$$

$$= \frac{12}{0.12} - \frac{(12)(2)}{(1+0.12/12)^{(12)(2)}-1} = 100 - 88.9763 = 11.0237$$

Appendix A gives $(A/G, 1\%, 24) = 11.0237$.

4.11 Mrs. Carter deposits \$100 in the bank at the end of each month. If the bank pays (*a*) 6% per year, (*b*) 7% per year, compounded monthly, how much money will she have accumulated at the end of 5 years?

(*a*) The effective monthly interest rate is $6\%/12 = 0.5\%$. There will be a total of $5 \times 12 = 60$ monthly payments. Hence, using Appendix A,

$$F = \$100(F/A, 0.5\%, 60) = \$100(69.7700) = \$6977.00$$

(*b*) The effective monthly interest rate is $7\%/12 = 0.583333\%$. As the tabulated value of $(F/A, 0.583333\%, 60)$ is not readily available, we interpolate linearly between $(F/A, 0.5\%, 60)$ and $(F/A, 0.75\%, 60)$.

$$(F/A, 0.583333\%, 60) \approx 69.7700 + \frac{0.583333 - 0.5}{0.75 - 0.5}(75.4241 - 69.7700) = 71.6547$$

and

$$F \approx \$100(71.6547) = \$7165.47$$

Another (and more accurate) way to solve this problem is to apply (*2.3*):

$$F \approx \$100 \frac{(1 + 0.00583333)^{60} - 1}{0.00583333} = \$100 \frac{0.417625}{0.00583333} = \$7159.29$$

4.12 In Problem 4.11, suppose that Mrs. Carter deposits \$100 a month during the first year, \$110 a month during the second year, \$120 a month during the third year, etc. How much will have accumulated at the end of 5 years if the interest rate is 6% per year, compounded monthly?

Treating each year separately,

$$\begin{aligned} F = {} & \$100(F/A, 0.5\%, 12)(F/P, 0.5\%, 48) + \$110(F/A, 0.5\%, 12)(F/P, 0.5\%, 36) \\ & + \$120(F/A, 0.5\%, 12)(F/P, 0.5\%, 24) + \$130(F/A, 0.5\%, 12)(F/P, 0.5\%, 12) \\ & + \$140(F/A, 0.5\%, 12) \end{aligned}$$

The required numerical values can be obtained from Appendix A (using interpolation in some cases), or from (*2.1*) and (*2.3*).

$$\begin{aligned} F &= [\$100(1.2705) + \$110(1.1967) + \$120(1.1272) + \$130(1.0617) + \$140](12.3356) \\ &= (\$671.972)(12.3356) = \$8289.18 \end{aligned}$$

Supplementary Problems

4.13 What is the effective annual interest rate if the nominal interest rate is 6%, compounded monthly?
Ans. 6.1678% per year

4.14 How many years will be required for a sum of money to double, if the annual interest rate is 10%, compounded quarterly? *Ans.* $n = 5.86$ years (6 years, if n must be an integer)

4.15 Mr. Smith plans to deposit money in a bank that pays 10% interest per year, compounded daily. What effective rate of interest will he receive (*a*) yearly? (*b*) semiannually?
Ans. (*a*) 10.515%; (*b*) 5.0625%

4.16 A bank pays interest at the rate of 12% per year, compounded monthly. If a man deposits \$3000 in the bank and leaves it for 5 years, how much money will accumulate, according to (*a*) (*2.1*)? (*b*) Appendix A? *Ans.* (*a*) \$5450.09; (*b*) \$5450.10

4.17 A person deposits \$2000 in a savings account. If all of the money is allowed to accumulate, how much will the person have at the end of 5 years, given a nominal interest rate of 6%, compounded (*a*) annually? (*b*) quarterly? (*c*) monthly? (*d*) daily? Use the tables in Appendix A to obtain the answers whenever possible. *Ans.* (*a*) \$2676.40; (*b*) \$2693.80; (*c*) \$2697.80; (*d*) \$2699.65

4.18 What amount of money is equivalent to receiving \$8000 three years from today, if the interest rate is 8% per year, compounded semiannually? *Ans.* \$6322.52

4.19 A bank pays 6% interest per year, compounded quarterly. To what amount will a \$5000 deposit grow if left in that bank for 10 years? *Ans.* \$9070.09

4.20 Repeat Problem 4.19 for annual compounding. *Ans.* \$8954.24

4.21 Calculate the amount of money that you would have in your savings account at the end of 12 months if you made the following deposits:

End of Month	1	3	6	7	8	11
Deposit, \$	200	90	70	75	85	70

Assume that the bank pays 6% interest per year, compounded semiannually, and that it pays simple interest on any interperiod deposits. *Ans.* \$611.73

4.22 Calculate the balance in Mr. Warren's account at the end of the year, if he deposits \$100 each at the ends of months 1 and 6, and \$200 each at the ends of months 7 and 9. His bank pays 8% per year, compounded quarterly, and simple interest on the interperiod deposits. *Ans.* \$622.30

4.23 Mr. Smith plans to deposit \$8000 in a savings account at the end of each year for 5 years. The bank pays interest at the rate of 12% per year, compounded quarterly, on such a plan. Calculate how much money Mr. Smith can expect to withdraw at the end of 5 years, (*a*) by the method of Example 4.4, (*b*) by use of (*4.3*). *Ans.* (*a*) \$51 382.40; (*b*) \$51 394.73

4.24 Suppose that \$2000 is invested now, \$2500 two years from now, and \$1200 four years from now, all at 8% per year, compounded quarterly. What will be the total amount 10 years from now?
Ans. \$11 057.33

4.25 A savings bank offers \$1000 certificates of deposit. Each certificate can be redeemed for \$2000 after $8\frac{1}{2}$ years. What is the nominal annual interest rate if the interest is compounded monthly? *Ans.* 8.182%

4.26 What is the effective annual interest rate for the certificates of Problem 4.25? *Ans.* 8.496%

4.27 Frank is trying to determine whether or not he can afford to borrow \$10 000 for 2 years. The bank charges 1% per month on the unpaid balance. Frank wants to repay the loan in 24 equal monthly installments, but feels he cannot pay more than \$450 per month. Can he afford the loan?
Ans. No (he would have to repay \$470.70 per month)

4.28 What will be the monthly payment on a 30-year, \$100 000 mortgage loan, where the interest rate is 12% per year, (*a*) compounded monthly? (*b*) compounded daily? *Ans.* (*a*) \$1028.61; (*b*) \$1033.25

4.29 What is the answer to Problem 4.28 under the approximation $\lim_{n\to\infty} (A/P, i, n) \approx i$? *Ans.* \$1000

4.30 Mrs. Jones plans to save \$750 a month for the next 10 years, at 10% per year, compounded monthly. How much money will she have at the end of 10 years? *Ans.* \$153 633.38

4.31 What is the present value of a series of monthly payments of \$300 each over 12 years, if the interest rate is 9% per year, compounded monthly? *Ans.* \$26 361.29

4.32 How much money must be deposited in a savings account each month to accumulate \$12 000 at the end of 5 years, if the bank pays interest at the rate of 6% per year, compounded monthly? *Ans.* \$171.99

4.33 Compute the amount of the monthly deposits Mr. Jones must make for the next 5 years in order for him to accumulate \$10 000 at the end of 5 years, at the nominal rate of 6% per year, compounded daily.
Ans. \$143.28

4.34 A series of quarterly payments of \$1000 for 25 years is economically equivalent to what present sum, if the quarterly payments are invested at an annual rate of 8%, compounded quarterly?
Ans. \$43 103.45

4.35 Using a suitable form of (*2.9*), evaluate A/G for $r = 12\%$, compounded monthly, and $n = 5$ years. Compare the result to the tabulated value in Appendix A. *Ans.* 26.5333 (same in Appendix A)

4.36 An investment plan pays 15% per year, compounded monthly. How much would have to be invested every year so that \$40 000 would be accumulated by the end of 10 years? *Ans.* \$1869.13

4.37 On the day of his son's birth, a father decided to establish a fund for the boy's college education. The father wants the son to be able to withdraw \$4000 from the fund on his 18th birthday, again on his 19th birthday, again on his 20th birthday, and again on his 21st birthday. If the fund earns interest at 9% per year, compounded quarterly, how much should the father deposit at the end of each year, up through the 17th year? Compare with the result obtained for annual compounding (Problem 2.25).
Ans. \$338.41

4.38 Solve Problem 4.37 again, assuming now that the money is deposited in the bank at the beginning, rather than the end, of each year. *Ans.* \$309.59

4.39 A new machine is expected to cost \$6000 and have a life of 5 years. Maintenance costs will be \$1500 the first year, \$1700 the second year, \$1900 the third year, \$2100 the fourth year, and \$2300 the fifth year. How much should be deposited in a fund that earns 9% per year, compounded monthly, in order to pay for this machine? *Ans.* \$13 180

4.40 Suppose that a person deposits \$2500 in a savings account at the end of each year for the next 15 years. If the bank pays (*a*) 8% per year, (*b*) $8\frac{1}{2}$% per year, compounded daily, how much money will the person have by the end of the 15th year? *Ans.* (*a*) \$69 609; (*b*) \$72 649

4.41 Jones has deposited his life savings of \$70 000 in a retirement income plan with a local bank. The bank pays (*a*) 10% per year, (*b*) 11.25% per year, compounded quarterly, on such deposits. What is the maximum fixed amount Jones can withdraw at the end of each year and still have the funds last for 15 years?
Ans. (*a*) \$9404.33; (*b*) \$10 132.00

4.42 Ann White is planning to take early retirement. She has decided that she needs \$15 000 a year for the first 5 years of retirement; after that, Social Security and other pension plans will provide her with adequate retirement income. How much money will she need to have in the bank at the start of the 5-year period, if the bank pays (*a*) 10% per year, (*b*) 6.75% per year, compounded monthly? Compare the result in (*a*) with Problem 2.31(*a*). *Ans.* (*a*) \$56 184; (*b*) \$61 565

4.43 Mr. Frank is planning for a 25-year retirement period, during which he wants to withdraw a portion of his savings at the end of each year. He plans to withdraw \$10 000 at the end of the first year, and to then increase the amount of the withdrawal by \$1000 each year (to offset inflation). How much money should he have in his savings account at the start of his retirement period in order to achieve these goals, if the bank pays 9% per year, compounded quarterly? Compare with Problem 2.32(*a*). *Ans.* \$169 740

4.44 Repeat Problem 4.43 for an interest rate of $8\frac{1}{2}$% per year, compounded quarterly. How significant is the $\frac{1}{2}$% difference? *Ans.* \$179 267 (the $\frac{1}{2}$% difference requires almost \$10 000 additional)

4.45 The cost to maintain a new car is estimated to be \$75 the first year, and to increase by \$12 each year thereafter. How much money should be set aside for maintenance, if the car is to be kept 6 years and if the money which is set aside earns interest at the rate of 5% per year, compounded monthly?
Ans. \$522.15

4.46 A father wants to set aside money for his 8-year-old son's college education, by making annual deposits to a bank account in his son's name that pays 8% per annum, compounded quarterly. What equal deposits must the father make on the son's 9th through 17th birthdays, in order for the son to be able to withdraw $4000 on each of his four birthdays from the 18th to the 21st? *Ans.* $1013.76

4.47 A savings account earns interest at the rate of $6\frac{3}{4}$% per year, compounded quarterly. How much money must initially be placed in the account to provide for fifteen end-of-year withdrawls, if the first withdrawal is $2000 and each subsequent withdrawal increases by $350? *Ans.* $36 792

4.48 A bank offers its customers a Christmas Club account in which they deposit $25 a week for 39 weeks, starting in February. At the end of the period (mid-November), each customer can withdraw $1000. What is the nominal annual interest rate, assuming monthly compounding? (*Hint*: The 39 weeks compose nine interest periods, at the ends of which the lumped deposits are $100, $100, $125, ...) *Ans.* 7.72%

4.49 Mr. Williams deposits $200 in the bank at the end of each quarter. If the bank pays 6% per year, compounded quarterly, how much money will Mr. Williams have accumulated at the end of 12 years? *Ans.* $13 913

4.50 Repeat Problem 4.49 for a nominal interest rate of $6\frac{1}{2}$% per year, compounded monthly. *Ans.* $14 373

4.51 Repeat Problem 4.49 for the case where the money is deposited at the *beginning* of each quarter. *Ans.* $14 122

4.52 An engineering student borrows $4000 to pay tuition for his senior year. Payments are to be made in 36 equal monthly installments, to begin the first month after graduation. How much money must the student repay each month, if he is graduated 9 months after taking out the loan and if the interest rate is 10% per year, compounded (*a*) monthly? (*b*) quarterly? (*c*) daily?
Ans. (*a*) $139.08; (*b*) $139.98; (*c*) $138.64

4.53 A recent engineering graduate intends to purchase a new car. He plans to pay $2000 down and to finance the balance over a 4-year period. The maximum amount that he can repay each month is $200. What is the most expensive car that he can afford, assuming an interest rate of 12% per year, compounded monthly? *Ans.* $9595

4.54 Suppose that the engineering graduate of Problem 4.53 can afford to repay $200 a month during the first year, $225 a month during the second year, $250 a month during the third year, and $275 a month during the fourth year. What is the most expensive car he can afford, assuming he pays $2000 down and the interest rate is 12% per year, compounded monthly? *Ans.* $10 878

4.55 Repeat Problem 4.54 for an interest rate of $10\frac{3}{4}$% per year, compounded daily. *Ans.* $11 094

4.56 A young couple are saving money in order to make a down payment on a house 6 years from now. Suppose that they save $150 a month during the first year, $165 a month during the second year, and so on, the amount increasing by $15 a month in each successive year. What is the most expensive house that they will be able to purchase at the end of the 6-year period, if they pay 25% down? Assume that their savings earns 7% per year, compounded quarterly. *Ans.* $65 327

4.57 An engineer plans to borrow $10 000 to open his own consulting business. He must repay $215 a month for 5 years. What is the nominal annual interest rate, based on monthly compounding?
Ans. 10.51%

Chapter 5

Continuous Compounding

5.1 NOMINAL AND EFFECTIVE INTEREST RATES

Continuous compounding can be thought of as a limiting case of the multiple-compounding situation of Section 4.3. Holding the nominal annual interest rate fixed at r and letting the number of interest periods become infinite, while the length of each interest period becomes infinitesimally small, we obtain from (*4.3*)

$$i = \lim_{m \to \infty} \left[\left(1 + \frac{r}{m} \right)^m - 1 \right] = e^r - 1 \tag{5.1}$$

as the expression for the effective annual interest rate in continuous compounding.

Example 5.1 A savings bank is selling long-term savings certificates that pay interest at the rate of $7\frac{1}{2}\%$ per year, compounded continuously. The bank claims that the actual annual yield of these certificates is 7.79%. What does this mean?

The nominal interest rate is $7\frac{1}{2}\%$. Since the interest is compounded continuously, the effective annual interest rate is given by (*5.1*) as

$$i = e^{0.075} - 1 = 0.077884 \approx 7.79\%$$

Formula (*5.1*) is very convenient, provided a calculator is available to carry out the exponentiation. Tabulated values of the effective annual interest rate may be used instead; see Appendix B.

5.2 DISCRETE PAYMENTS

If interest is compounded continuously but payments are made annually, we can still use the formulas of Chapter 2 for the various compound interest factors, provided i is given by (*5.1*). Thus:

$$F/P = e^{rn} \tag{5.2}$$

$$P/F = e^{-rn} \tag{5.3}$$

$$F/A = \frac{e^{rn} - 1}{e^r - 1} \tag{5.4}$$

$$A/F = \frac{e^r - 1}{e^{rn} - 1} \tag{5.5}$$

$$A/P = \frac{e^r - 1}{1 - e^{-rn}} \tag{5.6}$$

$$P/A = \frac{1 - e^{-rn}}{e^r - 1} \tag{5.7}$$

$$A/G = \frac{1}{e^r - 1} - \frac{n}{e^{rn} - 1} \tag{5.8}$$

where n represents the number of years, as before. These factors are denoted $[F/P, r\%, n]$, $[P/F, r\%, n]$, etc. (Notice the use of square brackets rather than parentheses, and reference to the nominal interest rate, to indicate continuous compounding.) The continuous compound interest factors can be evaluated directly from the above formulas, or they can be obtained from the tables presented in Appendix C.

Example 5.2 A savings bank offers long-term savings certificates at $7\frac{1}{2}\%$ per year, compounded continuously. If a 10-year certificate costs \$1000, what will be its value at maturity? Compare with the value that would be obtained if the interest were compounded annually rather than continuously.

From (*5.2*),

$$F = P \times [F/P, r\%, n] = \$1000\, e^{(0.075)(10)} = \$2117.00$$

This problem can also be solved using Appendix C. Since a table is not available for a nominal interest rate of $7\frac{1}{2}\%$ per year, however, it will be necessary to interpolate between the 7% and 8% values.

$$[F/P, 7\%, 10] = 2.0138 \qquad [F/P, 8\%, 10] = 2.2255$$

and
$$[F/P, 7\tfrac{1}{2}\%, 10] \approx 2.0138 + \frac{7.5 - 7.0}{8.0 - 7.0}(2.2255 - 2.0138) = 2.1197$$

The future worth of the savings certificate can now be obtained as

$$F \approx \$1000(2.1197) = \$2119.70$$

If the interest were compounded annually rather than continuously, the future worth would be

$$F = \$1000(1 + 0.075)^{10} = \$2061.00$$

or \$56 less than the amount that is obtained with continuous compounding.

Example 5.3 A savings account earns interest at the rate of 6% per year, compounded continuously. How much money must initially be placed in the account to provide for twenty end-of-year withdrawals, if the first withdrawal is \$1000 and each subsequent withdrawal increases by \$200?

The solution may be formulated after (*2.11*):

$$P = \$1000[P/A, 6\%, 20] + \$200[A/G, 6\%, 20]\,[P/A, 6\%, 20]$$

From (*5.7*) and (*5.8*),

$$[P/A, 6\%, 20] = \frac{1 - e^{-(0.06)(20)}}{e^{0.06} - 1} = 11.3009$$

$$[A/G, 6\%, 20] = \frac{1}{e^{0.06} - 1} - \frac{20}{e^{(0.06)(20)} - 1} = 7.5514$$

(these values can also be obtained from Appendix C); hence,

$$P = \$1000(11.3009) + \$200(7.5514)(11.3009) = \$28\,368.42$$

If interest is compounded continuously but payments are made p times a year, formulas (*5.2*) through (*5.8*) remain valid with r replaced by r/p and with n replaced by np. [These substitutions do not, of course, alter the forms of (*5.2*) and (*5.3*).]

Example 5.4 A person borrows \$5000 for 3 years, to be repaid in 36 equal monthly installments. The interest rate is 10% per year, compounded continuously. How much money must be repaid at the end of each month?

Calculating $[A/P, 10\%/12, 36]$ by (*5.6*), we have

$$A = \$5000\,\frac{e^{0.10/12} - 1}{1 - e^{-0.30}} = \$161.43$$

Example 5.5 A bank offers its customers a Christmas Club account, in which they deposit \$12.61 a week for 39 weeks, starting in mid-February. At the end of 39 weeks (mid-November), each customer will have accumulated \$500, which can be withdrawn to pay for gifts and other seasonal expenses. What is the nominal interest rate, assuming continuous compounding?

We know that $F/A = \$500/\$12.61 = 39.6511$. Let us attempt to choose an interest rate that will yield this value when substituted into (*5.4*), which in this case takes the form

$$F/A = \frac{e^{rn} - 1}{e^{r/p} - 1}$$

with $n = 0.75$ year and $p = 52$ payment periods per year:

$$\textit{for } r = 3\% \qquad F/A = \frac{e^{(0.03)(0.75)} - 1}{e^{0.03/52} - 1} = 39.4308$$

$$\textit{for } r = 4\% \qquad F/A = \frac{e^{(0.04)(0.75)} - 1}{e^{0.04/52} - 1} = 39.5757$$

$$\textit{for } r = 5\% \qquad F/A = \frac{e^{(0.05)(0.75)} - 1}{e^{0.05/52} - 1} = 39.7214$$

The desired value, $F/A = 39.6511$, lies somewhere between the 4% and the 5% values. Thus, using linear interpolation,

$$r \approx 4 + \frac{39.6511 - 39.5757}{39.7214 - 39.5757}(5 - 4) = 4.52\%$$

5.3 CONTINUOUS PAYMENTS

For p payments of A per year and continuous compounding, we have, as in Example 5.5,

$$F = A\,\frac{e^{rn} - 1}{e^{r/p} - 1} = \frac{\bar{A}(e^{rn} - 1)}{p(e^{r/p} - 1)}$$

where $\bar{A} \equiv A/p^{-1}$ is the average rate of payment over a payment period. With $\bar{A}$ held fixed, the denominator above approaches r as $p \to \infty$; and we obtain for *continuous payments at rate* $\bar{A}$ (dollars per unit time)

$$F/\bar{A} = \frac{e^{rn} - 1}{r} \tag{5.9}$$

In like manner, we find:

$$\bar{A}/F = \frac{r}{e^{rn} - 1} \tag{5.10}$$

$$\bar{A}/P = \frac{re^{rn}}{e^{rn} - 1} \tag{5.11}$$

$$P/\bar{A} = \frac{e^{rn} - 1}{re^{rn}} \tag{5.12}$$

The notation $[F/\bar{A}, r\%, n]$, etc., is used for these factors.

Example 5.6 At what rate must funds be continuously added to a savings account in order to accumulate \$10 000 in 15 years, if interest is paid at 5% per year, compounded continuously?

By (*5.10*),

$$\bar{A} = \$10\,000\,\frac{0.05}{e^{(0.05)(15)} - 1} = \$447.63 \text{ per year}$$

that is, \$447.63 must flow uniformly into the account each year.

It is interesting to compare this result to a series of uniform, end-of-year payments, with interest compounded continuously as above. The amount of each such payment is given by (*5.5*) as

$$A = \$10\,000[A/F, 5\%, 15] = \$10\,000(0.0459) = \$459 \text{ per year}$$

Thus, an additional \$11.37 would be required each year if the payments were made annually rather than continuously.

It is convenient to determine the end-of-year payment A equivalent to continuous payments at rate $\bar{A}$, under continuous compounding. If (*5.4*) and (*5.9*) are to give the same value of F,

$$F = A\,\frac{e^{rn} - 1}{e^{r} - 1} = \bar{A}\,\frac{e^{rn} - 1}{r}$$

and so

$$A/\bar{A} = \frac{e^{r} - 1}{r} \tag{5.13}$$

This factor, written symbolically as $[A/\bar{A}, r\%]$, is tabulated in Appendix D.

Example 5.7 Rework Example 5.6, using the concept of the equivalent yearly payment.

The solution can be formulated as

$$\bar{A} = F[A/F, 5\%, 15]\,[\bar{A}/A, 5\%] = \frac{F\,[A/F, 5\%, 15]}{[A/\bar{A}, 5\%]}$$

From Appendixes C and D, $[A/F, 5\%, 15] = 0.0459$ and $[A/\bar{A}, 5\%] = 1.0254$. Thus,

$$\bar{A} = \frac{(\$10\,000)(0.0459)}{1.0254} = \$447.63 \text{ per year}$$

Solved Problems

5.1 What effective annual interest rate corresponds to a nominal interest rate of 10% per year, compounded continuously?

$$i = e^{0.10} - 1 = 0.105171 \approx 10.52\%$$

(This result could also have been obtained from Appendix B.)

5.2 Determine the nominal interest rate corresponding to an effective interest rate of 10% per year, compounded continuously.

Solving (*5.1*) by natural logarithms,

$$r = \ln(1 + i) = \ln 1.10 = 0.0953$$

or 9.53%.

5.3 How much money must be deposited in a savings account so that \$5500 can be withdrawn 12 years hence, if the interest rate is 9% per year, compounded continuously, and if all the interest is allowed to accumulate? Compare the answer with the result obtained for annual compounding, in Problem 2.2.

$$P = \$5500[P/F, 9\%, 12] = \$5500[F/P, 9\%, 12]^{-1} = \$5500(2.9447)^{-1} = \$1867.76$$

Comparing with Problem 2.2, we see that a savings of \$87.66 is realized by continuous compounding.

5.4 How much money must be deposited at the end of each year in a savings account that pays 9% per year, compounded continuously, to have a total of \$10 000 at the end of 14 years? Compare the answer with the result obtained for annual compounding, in Problem 2.6.

$$A = \$10\,000[A/F, 9\%, 14] = \$10\,000[F/A, 9\%, 14]^{-1} = \$10\,000(26.8165)^{-1} = \$372.90$$

If the interest were compounded annually rather than continuously, the yearly deposit would have to be \$11.43 greater.

5.5 Mr. Smith is planning his retirement. He has decided that he will need \$12 000 per year to live on, in addition to his other retirement income from Social Security and a private pension plan. How much money should he plan to have in the bank at the start of his retirement, if the bank pays 10% per year, compounded continuously, and if Mr. Smith wants to make 12 annual withdrawals of \$12 000 each?

$$P = A\,[P/A, 10\%, 12] = A\,[A/P, 10\%, 12]^{-1} = (\$12\,000)(0.15050)^{-1} = \$79\,734.22$$

In Problem 2.9 we saw that the required amount of money would be \$81 766.15 if the interest were compounded annually rather than continuously. Hence, the continuous compounding results in a savings of over \$2000.

5.6 Ms. Brown deposits \$1000 in the bank at the end of the first year, \$1200 at the end of the second year, etc., continuing to increase the amount by \$200 a year, for 20 years. If the bank pays 7% per year, compounded continuously, how much money will have accumulated at the end of 20 years?

Writing $A' = \$1000 + A$, we have, from Section 2.7,

$$F = A'\,[F/A', 7\%, 20] = \{\$1000 + G\,[A/G, 7\%, 20]\}\,[F/A', 7\%, 20]$$

Substituting $G = \$200$ and using Appendix C,

$$F = [\$1000 + \$200(7.2453)](42.1359) = \$103\,193.35$$

5.7 Calculate the factors $[F/A, 7\%, 20]$ and $[A/G, 7\%, 20]$ used in Problem 5.6.

By (*5.4*),

$$[F/A, 7\%, 20] = \frac{e^{(0.07)(20)} - 1}{e^{0.07} - 1} = \frac{4.055200 - 1}{1.072508 - 1} = 42.1359$$

and, by (*5.8*),

$$[A/G, 7\%, 20] = \frac{1}{e^{0.07} - 1} - \frac{20}{e^{(0.07)(20)} - 1} = \frac{1}{1.072508 - 1} - \frac{20}{4.055200 - 1} = 7.2454$$

5.8 Mrs. Carter deposits \$100 in the bank at the end of each month. If the bank pays (*a*) 6% per year, (*b*) 7% per year, compounded continuously, how much money will she have accumulated at the end of 5 years? (Compare Problem 4.11.)

(*a*) The nominal monthly interest rate is $6\%/12 = 0.5\%$. There will be a total of $5 \times 12 = 60$ monthly payments. Hence,

$$F = \$100[F/A, 0.5\%, 60]$$

From Appendix C, $[F/A, 0.5\%, 60] = 69.7970$; therefore,

$$F = \$100(69.7970) = \$6979.70$$

(*b*) The nominal monthly interest rate is $7\%/12 = 0.583333\%$. As a tabulated value of $[F/A, 0.583333\%, 60]$ is not available, we interpolate linearly between $[F/A, 0.5\%, 60]$ and $[F/A, 0.75\%, 60]$:

$$[F/A, 0.583333\%, 60] \approx 69.7970 + \frac{0.583333 - 0.5}{0.75 - 0.5}(75.4912 - 69.7970) = 71.6951$$

The desired solution is then $F \approx \$100(71.6951) = \$7169.51$.

A more accurate procedure would be to use (*5.4*), with r replaced by $r/12$:

$$F = \$100\,\frac{e^{(0.07)(5)} - 1}{e^{0.07/12} - 1} = \$100\,\frac{1.419068 - 1}{1.005850 - 1} = \$7163.56$$

5.9 In Problem 5.8, suppose Mrs. Carter deposits \$100 a month during the first year, \$110 a month during the second year, \$120 a month during the third year, etc. How much will have accumulated at the end of 5 years if the interest rate is 6% per year, compounded continuously? (Compare Problem 4.12.)

Proceeding as in Problem 4.12, we have:

$$\begin{aligned} F = {} & \$100[F/A, 0.5\%, 12]\,[F/P, 6\%, 4] + \$110[F/A, 0.5\%, 12]\,[F/P, 6\%, 3] \\ & + \$120[F/A, 0.5\%, 12]\,[F/P, 6\%, 2] + \$130[F/A, 0.5\%, 12]\,[F/P, 6\%, 1] \\ & + \$140[F/A, 0.5\%, 12] \end{aligned}$$

The required numerical values can be obtained from Appendix C. Thus,

$$F = [\$100(1.2712) + \$110(1.1972) + \$120(1.1275) + \$130(1.0618) + \$140](12.3364)$$
$$= (\$672.146)(12.3364) = \$8291.86$$

A neater solution procedure (which might also have been applied in Problem 4.12) is to consider the deposits as constituting a five-year gradient series, with

$$A_0 = \$100[F/A, 0.5\%, 12] \qquad \text{and} \qquad G = \$10[F/A, 0.5\%, 12]$$

Thus, as in Problem 5.6,

$$F = \{\$100[F/A, 0.5\%, 12] + \$10[F/A, 0.5\%, 12]\,[A/G, 6\%, 5]\}\,[F/A', 6\%, 5]$$
$$= \{\$100 + \$10[A/G, 6\%, 5]\}\,[F/A, 0.5\%, 12]\,[F/A', 6\%, 5]$$
$$= [\$100 + \$10(1.8802)](12.3364)(5.6578) = \$8291.87$$

5.10 Suppose that \$2000 is deposited each year, on a continuous basis, into a savings account that pays 6% per year, compounded continuously. How much money will have accumulated after 12 years?

With $\bar{A} = \$2000$ per year, (*5.9*) gives

$$F = \bar{A}\,\frac{e^{rn} - 1}{r} = \$2000\,\frac{e^{(0.06)(12)} - 1}{0.06} = \$2000(17.5739) = \$35\,147.77$$

Alternatively, using the tabular values in Appendixes C and D,

$$F = \$2000[A/\bar{A}, 6\%]\,[F/A, 6\%, 12] = \$2000(1.030609)(17.0519) = \$35\,147.68$$

Supplementary Problems

5.11 What effective annual interest rate corresponds to a nominal interest rate of 15% per year, compounded continuously? *Ans.* 16.18%

5.12 What nominal interest rate corresponds to an effective interest rate of 12% per year, compounded continuously? *Ans.* 11.33%

5.13 Determine the effective annual interest rate corresponding to a nominal interest rate of $8\frac{1}{2}\%$ per year, if the interest is compounded (*a*) quarterly, (*b*) monthly, (*c*) daily, (*d*) continuously.
Ans. (*a*) 8.77%; (*b*) 8.84%; (*c*) 8.871%; (*d*) 8.872%

5.14 An investment plan pays 15% per year, compounded continuously. How much would have to be invested at the end of each year so that \$40 000 will be accumulated by the end of 10 years? Compare with the result obtained for annual compounding (Problem 2.20). *Ans.* \$1859.26

5.15 Suppose that \$2000 is invested now, \$2500 is invested two years from now, and \$1200 is invested four years from now, all at 8% per year, compounded continuously. What will be the total amount 10 years from now? *Ans.* \$11 131.57

5.16 Repeat Problem 2.25 for continuous compounding. *Ans.* \$334.23

5.17 Repeat Problem 2.27(*a*) for continuous compounding. *Ans.* \$13 173.00

5.18 Repeat Problem 2.29(*a*) for continuous compounding. *Ans.* \$69 642.25

5.19 Repeat Problem 2.30 for continuous compounding. *Ans.* (*a*) \$9476.60; (*b*) \$10 226.83

5.20 Repeat Problem 2.31(*a*) for continuous compounding. *Ans.* $56 118.82

5.21 Repeat Problem 2.32(*a*) for continuous compounding. *Ans.* $167 884.49

5.22 Repeat Problem 2.32 for continuous compounding at $8\frac{1}{2}$%. How significant is the $\frac{1}{2}$% difference vis-a-vis Problem 5.21? *Ans.* $177 478.94

5.23 The cost of maintaining a new car is estimated to be $75 the first year and to increase by $12 each year thereafter. How much money should be set aside for maintenance, if the car is to be kept 6 years and if the money which is set aside earns interest at the rate of 5% per year, compounded continuously? (Compare Problem 4.45.) *Ans.* $521.95

5.24 Rework Problem 4.46 for continuous compounding. *Ans.* $1001.16

5.25 Find the monthly payment on a 30-year, $100 000 mortgage loan, where the interest rate is 12% per year, compounded continuously. *Ans.* $1033.25 [cf. Problem 4.28(*b*)]

5.26 Repeat Problem 4.30 for continuous compounding. *Ans.* $154 001.91

5.27 Repeat Problem 4.31 for continuous compounding. *Ans.* $26 317.24

5.28 Repeat Problem 4.32 for continuous compounding. *Ans.* $171.93

5.29 Mr. Smith plans to purchase a new $10 000 automobile. He wants to borrow all the money for the car, and repay it in equal monthly installments over a 4-year period. The nominal interest rate is 11% per year, compounded continuously. What will be Mr. Smith's monthly payment? *Ans.* $258.70

5.30 Repeat Problem 4.33 for continuous compounding. *Ans.* $143.27

5.31 A savings bank offers $1000 certificates of deposit. Each certificate can be redeemed for $2000 after $8\frac{1}{2}$ years. What are (*a*) the nominal, (*b*) the effective, annual interest rate, if the interest is compounded continuously? *Ans.* (*a*) 8.155%; (*b*) 8.496%

5.32 A savings account earns interest at the rate of $6\frac{3}{4}$% per year, compounded continuously. How much money must initially be placed in the account to provide for 15 end-of-year withdrawals if the first withdrawal is $2000 and each subsequent withdrawal increases by $350? *Ans.* $36 620.17

5.33 Repeat Problem 4.48 for continuous compounding. *Ans.* 6.90%

5.34 Repeat Problem 4.49 for continuous compounding. *Ans.* $13 953.93

5.35 Repeat Problem 4.50 for continuous compounding. *Ans.* $14 423.37

5.36 Repeat Problem 4.51 for continuous compounding. *Ans.* $14 163.24

5.37 A savings account pays $5\frac{1}{2}$% per year, compounded continuously. How much money must be deposited at the end of each month in order to accumulate $10 000 at the end of 7 years? *Ans.* $97.82

5.38 Repeat Problem 4.52 for continuous compounding. *Ans.* $139.21

5.39 Repeat Problem 4.53 for continuous compounding. *Ans.* $9586.27

5.40 Repeat Problem 4.54 for continuous compounding. *Ans.* $10 867.03

5.41 Repeat Problem 4.54 for continuous compounding at $10\frac{3}{4}$% per annum. *Ans.* $11 094.03

5.42 Repeat Problem 4.57 for continuous compounding. *Ans.* $65 824.16

5.43 Repeat Problem 4.57 for continuous compounding at $6\frac{1}{2}\%$ per year. *Ans.* \$64 878.08

5.44 An engineer borrows \$10 000 to buy a personal computer. He must repay \$218.94 a month for 5 years. What is the nominal annual interest rate, based upon continuous compounding? *Ans.* $11\frac{1}{4}\%$

5.45 A consulting firm has a continuous cash inflow of \$5 million a year. If this money is accumulated in an account that earns (*a*) 15% per year, (*b*) 13.5% per year, compounded continuously, how much money will have accumulated after 7 years? *Ans.* (*a*) \$61 921 704; (*b*) \$58 252 347

5.46 A company has set aside \$10 million to promote a new product. The money is to be spent continuously over a 3-year period (during which it is assumed that the sales of the product will offset expenses). If the \$10 million is placed in an account that earns (*a*) 12% per year, (*b*) 10.75% per year, compounded continuously, what is the maximum rate at which money can be withdrawn during the 3-year period?
Ans. (*a*) \$3 969 256 per year; (*b*) \$3 899 674 per year

5.47 In Problem 5.46, how much money must be placed in the account if \$4 million is to be spent on a continuous basis each year, and the interest rate is 10% per year, compounded continuously?
Ans. \$10 367 271

Chapter 6

Equivalence

6.1 ECONOMIC EQUIVALENCE

In economic analysis, "equivalence" means "the state of being equal in value." The concept is primarily applied in the comparison of different cash flows. As we know from earlier chapters, money changes value with time; therefore, one of the main factors when considering equivalence is to determine at which point(s) in time the money transactions occur. A second factor is the specific amounts of money involved in the transactions. Finally, the interest rate at which the equivalence is evaluated must also be considered.

Example 6.1 Bob, an engineering student, has just received his salary for a summer job. After living expenses and entertainment, he has left $1000, which he plans to save for a down payment on a new car. His father wants to borrow Bob's $1000, and promises to return $1060 one year from now. According to his father, that is what Bob would receive if he put the money in his bank savings account, which pays an effective annual interest rate of 6%. What should Bob do?

If Bob's only alternatives are lending the money or depositing it in his current savings account, both courses of action are indeed equivalent. That is, either would provide Bob, one year from now, with $1060 in return for his decision to forego using his $1000 today. Given this equivalence, Bob's decision would be based on factors external to engineering economics (e.g., the degree to which he trusts his father).

However, if Bob had a different savings option—say, a savings certificate with a guaranteed 9% annual yield—the equivalent value of his assets one year from now would be $1090. In this case, the lending and savings alternatives are no longer equivalent, and Bob has the problem of explaining this to his father.

In Example 6.1, $(F/P, 6\%, 1)$ served as the *equivalencing factor.* In general, all the compounding and discounting factors presented in earlier chapters are equivalencing factors.

Equivalence is not always directly apparent. Cash flows that have very different structures (i.e., different amounts being transacted at different points in time) may be equivalent at a certain interest rate.

Example 6.2 A company which produces and markets microcomputers has just introduced a new line which is expected to sell for $10 000 per system. Owing to market conditions, the company is being forced to offer financial incentives to potential customers. The company has decided to charge an interest rate of 10%, compounded yearly, and to give customers three options.

***Option 1*:** Pay in four equal yearly installments of

$$A = \$10\,000(A/P, 10\%, 4) = \$10\,000(0.3155) = \$3155$$

***Option 2*:** Pay the interest each year, and the principal (and interest) at the end of the fourth year. This means paying $1000 ($\$10\,000 \times 0.10$) at the end of years 1, 2, and 3, and $11 000 at the end of year 4.

***Option 3*:** Make a single payment of

$$F = \$10\,000(F/P, 10\%, 4) = \$10\,000(1.4641) = \$14\,641$$

at the end of year 4.

Which option is best for a customer? for the company?

As summarized in Table 6-1, the three payment plans offered a customer are quite different in structure. However, *if* 10% is the "appropriate" interest rate for his economic evaluations, all three plans are equivalent: each provides him with a microcomputer worth $10 000 and gives him 4 years to repay at a 10% interest rate, compounded yearly. From the company's point of view, similar reasoning applies: at 10% interest, all plans are equivalent because all result in the sale of a $10 000 item, and the money is recovered over a 4-year period.

Table 6-1

End of Year	Payment		
	Option 1	Option 2	Option 3
1	$3155	$ 1 000	$ 0
2	3155	1 000	0
3	3155	1 000	0
4	3155	11 000	14 641

Notice that different cash flows are equivalent if they have the same value at some point in time.

Example 6.3 Are the financing plans of Example 6.2 still equivalent if the evaluation is made at the end of year 4?

Yes; at the end of year 4, all plans have an equivalent value of $14 641 (up to roundoff errors) when the same interest rate (10%) is used to make the evaluations.

Option 1 (equal payments):

$$F = \$3155(F/A, 10\%, 4) = \$3155(4.641) = \$14\,642$$

Option 2 (amortization of interest):

$$F = \$1000(F/A, 10\%, 4) + \$10\,000 = \$14\,641$$

More generally, we can say that options 1 and 3 must be equivalent at the end of year 4, since they are obviously equivalent at the end of year 0; and that options 2 and 3 are equivalent, on the basis of the relation

$$(F/P, i\%, n) = i\,(F/A, i\%, n) + 1$$

derived in Problem 3.3(*a*). Thus, all three options are equivalent, to the customer and to the company, *provided the two parties use the same interest rate.*

Example 6.4 The company in Example 6.2 still uses interest rate $i = 10\%$, and therefore still offers the same three payment plans. The customer, however, calculates interest at rate i', so that the values (costs) to him of the three options at the end of year 0 are as given in Table 6-2; for instance, for $i' = 12\%$, the value of option 2 is

$$P = \$1000(P/A, 12\%, 4) + \$10\,000(P/F, 12\%, 4)$$
$$= \$1000(3.0374) + \$10\,000(0.6355) = \$9392$$

Table 6-2

Customer's Rate, i'	Present Value		
	Option 1	Option 2	Option 3
8%	$10 450	$10 662	$10 760
10%	10 000	10 000	10 000
12%	9 583	9 392	9 304

Note that when $i' \neq 10\%$, the options are no longer equivalent to the customer, and, more important, different interest rates may lead to different decisions. Consider, for example, a customer who has the money ($10 000) on hand, but who knows that he can put that money in a savings account which pays 12% effective interest, compounded yearly. The best strategy for this customer would be to take option 3 and pay $14 641 four years from now. Since this amount is equivalent to only $9304 at the rate he has saved his money, he would end up with a net saving of $10 000 − $9304 = $696 (year 0 money). On the other hand, if a customer is used to paying only 8% for loans, his best alternative (assuming he cannot pay cash or borrow at that rate to buy the computer) is option 1, the least costly at the interest rate he normally uses to make his economic evaluations.

6.2 THE COST OF CAPITAL

From Example 6.4 it is seen that the relative evaluation of cash flows depends critically on the "appropriate" or "pertinent" interest rate used in the calculations. Unfortunately, the interest rate that determines the time value of money is not usually known, nor is it easy to determine. It stands to reason, though, that if money is to be invested in a project, the project's cash flow equivalent value should be calculated at an interest rate that exceeds the rate incurred in raising the initial capital. The extra percentage points are justified in terms of risks associated with the specific project and with long-term commitment of funds, and in terms of a profit margin required to get involved in the economic activity. Thus, a mining company considering diversification into plastics would use a higher interest rate to evaluate such a project than it would use for a new mining project, because of the risk of entering a new venture with unknown market factors and for which no experience is available.

There are several means for a company to raise money for a project. It may borrow from a bank at a specified interest rate; it may reinvest profits from other projects instead of distributing them to the owners or shareholders; it may sell stock, thereby increasing the number of owners (and decreasing the equity of each stockholder); and it may borrow from the public through the issue of bonds. Almost always, a combination of methods is employed, and one way to measure the cost of capital is to calculate a weighted average of the costs of funds acquired from all sources.

6.3 STOCK VALUATION

Stock represents a share of ownership in a company. Its equivalent-value calculation presents practical difficulties of estimating future dividends and selling price, which are affected not only by the company's performance but also by the overall situation of the economy and of the stock market.

Example 6.5 ABC Corporation's stock, which currently sells for $50 per share, has been paying a $3 annual dividend per share and increasing in value at an average rate of 5% per year, over the last 5 years. It is expected that the company's stock will maintain this performance over the next 5 years. (*a*) What is the company's cost of the capital raised through the selling of this stock? (*b*) Is this stock a good buy for an investor who expects a 9% return on his investments?

(*a*) From the company's point of view, it will receive $50 per share today and would have to pay

$$\$50(1.05)^5 = \$63.81$$

to buy it back 5 years from now. In addition, it must pay a yearly dividend of $3 per share. The equation of value at time 0 for this cash flow is therefore

$$\$50 = \$3(P/A, i\%, 5) + \$63.81(P/F, i\%, 5)$$

where $i\%$ is the cost of capital for money raised through the sale of this stock, assuming the forecast dividends and selling price are accurate. The solution for i must be found by a trial-and-error approach:

$$\textit{for } i = 10\% \qquad \$3(3.7907) + \$63.81(0.62092) = \$50.99$$
$$\textit{for } i = 11\% \qquad \$3(3.6958) + \$63.81(0.59345) = \$48.96$$

Thus, by linear interpolation,

$$i \approx 10\% + (11\% - 10\%)\frac{50.99 - 50.00}{50.99 - 48.96} = 10.49\%$$

(*b*) For a customer who wants to make 9% on his investments, the value of his expected receipts from this share is given by

$$P = \$3(P/A, 9\%, 5) + \$63.81(P/F, 9\%, 5) = \$3(3.8896) + \$63.81(0.64993) = \$53.14$$

Since the expected value of the share exceeds the asked price, it is expedient for him to buy it; if he does so, he should not only recover his $50 investment and the 9% he expects yearly, but he should gain an extra $3.14 (year 0 money) in the transaction.

Alternatively, since we infer from (*a*) that the company expects the stock to yield 10.49% to an investor, the investor ought to buy it, if he requires only 9% and if he agrees with the company as to the stock's future performance.

6.4 BOND VALUATION

A *bond* is an economic instrument which has a *face value* guaranteed to be paid to the bondholder by the issuing company when the instrument reaches maturity. In addition, the bondholder usually receives periodic dividends at a specified interest rate. Bonds are transacted on the market, and their value depends on the size and timing of the dividends, the duration before maturity, and the rate of return desired by the bond purchaser. The company's cost of the capital raised through bonds will depend on their acceptability to the public.

Example 6.6 ABC Corporation has decided to sell $1000 bonds which will pay semiannual dividends of $20 (2% per period) and will mature in 5 years. The bonds are sold at $830, but after brokers' fees and other expenses the company ends up receiving $760. (*a*) What is the company's cost of the capital raised through the sale of these bonds? (*b*) Is the bond a good buy for an investor who expects a 9% return on his investments?

(*a*) From the company's point of view, it will receive $760 per bond today and will have to pay $1000 (the face value) 5 years hence, plus a $20 semiannual dividend. The equation of value at time 0 for this cash flow is

$$\$760 = \$20(P/A, i\%, 10) + \$1000(P/F, i\%, 10)$$

where 10 periods are used because dividends are paid twice a year and the bond matures in 5 years, and where $i\%$ is the cost of capital, effective per 6-month period. Solving by trial and error:

$$\textbf{for } i = 5\% \qquad \$20(7.7216) + \$1000(0.61392) = \$768.35$$
$$\textbf{for } i = 6\% \qquad \$20(8.1108) + \$1000(0.55840) = \$720.62$$

and linear interpolation gives

$$i \approx 5\% + \frac{768.35 - 760.00}{768.35 - 720.62}(6\% - 5\%) = 5.17\%$$

The company's cost of capital for money raised through the sale of these bonds is, on a yearly basis, given by (*4.3*) as

$$i = (1.0517)^2 - 1 = 10.62\%$$

(*b*) From the investor's point of view, he will pay $830 today to receive $20 every 6 months and $1000 in 5 years. He expects an effective rate of 9% a year on his investments, or

$$\sqrt{1.09} - 1 = 4.40\%$$

per 6-month period. Hence the equivalent value at time 0 of his expected receipts is

$$\begin{aligned} P &= \$20(P/A, 4.40\%, 10) + \$1000(P/F, 4.40\%, 10) \\ &= \$20\frac{1-(1.0440)^{-10}}{0.0440} + \$1000(1.0440)^{-10} = \$809.12 \end{aligned}$$

This value represents the maximum amount the investor can bid for this bond, if he requires an effective annual yield of 9% on his investments. Since the market value today ($830) exceeds this maximum amount, he should look for another business opportunity which could give him his required 9% return.

Notice that we could *not* conclude from (*a*) that the investor could realize 10.62% (>9%); for, in effect, part of that 10.62% goes to the brokers.

6.5 MINIMUM ATTRACTIVE RATE OF RETURN

If, as is often the case, the interest rate at which a project should be evaluated is not known, a *target rate, cut-off rate,* or *valuation rate* will be used. This rate is also called the *minimum attractive rate of return* (abbreviated MARR). While dependent on general company policy, the MARR may also be project specific, and will normally increase with the risk attending the project. It will certainly be higher than the cost of raising capital for the project, estimated as described in Section 6.2.

Example 6.7 In order to finance a $100 000 project, ABC Corporation has decided to raise $20 000 through the sale of stock, as described in Example 6.5, and $30 000 through the issuance of bonds, as described in Example 6.6. For the balance, $10 000 will be borrowed from a bank at an annual interest rate of 12% and $40 000 will be reinvested from last year's profits. (*a*) What is the project's cost of capital? (*b*) What MARR should be used to evaluate this project?

(*a*) For the four sources of capital, we have:

$$\textbf{\textit{stock}} \qquad \text{weight} = \frac{20\,000}{100\,000} \qquad \text{cost} = 10.49\%$$

$$\textbf{\textit{bonds}} \qquad \text{weight} = \frac{30\,000}{100\,000} \qquad \text{cost} = 10.62\%$$

$$\textbf{\textit{loan}} \qquad \text{weight} = \frac{10\,000}{100\,000} \qquad \text{cost} = 12.00\%$$

$$\textbf{\textit{reinvestment}} \qquad \text{weight} = \frac{40\,000}{100\,000} \qquad \text{cost} = 10.49\%$$

and so

$$\text{weighted average cost} = \sum_{\substack{\text{all} \\ \text{sources}}} (\text{source weight})\,(\text{source cost})$$

$$= (0.2)\,(10.49) + (0.3)\,(10.62) + (0.1)\,(12.00) + (0.4)\,(10.49) = 10.68\%$$

Note that the cost of capital for reinvested profits was assumed to be equal to the cost of common stock: this is a minimal value, based on the consideration that stockholders are being denied dividends from the retained earnings.

(*b*) The MARR for this project must exceed 10.68%.

The MARR will be treated in further detail in Chapter 9.

6.6 FAIR MARKET VALUE

The concept of equivalence, as applied in the foregoing examples, may be used to determine the actual cost of a loan, the maximum amount a person or company can bid on a desired property or equipment, and, in general, in the determination of the "fair market value" of an asset.

Example 6.8 An engineering firm has turned to Friendly Shark, Inc., to borrow $30 000 needed for a short-term (2-year) project, attracted by an advertisement announcing an interest rate of 12% per year. Friendly Shark's loan statement indicates the following:

Interest: ($30 000) (1% per month) (24 months) =	$ 7 200
Loan	30 000
Total	$37 200
Monthly installment = $37 200/24 = $1550	

What is the actual cost of borrowing money from Friendly Shark, Inc.?

The engineering firm receives $30 000 immediately and must pay back $1550 per month over a 24-month period. The monthly interest rate i which makes these flows equivalent satisfies

$$\$30\,000 = \$1550(P/A, i\%, 24) \qquad \text{or} \qquad (P/A, i\%, 24) = 19.355$$

Now, $(P/A, 1.5\%, 24) = 20.030$ and $(P/A, 2.0\%, 24) = 18.914$. Hence, by interpolation,

$$i \approx 1.5\% + \frac{20.030 - 19.355}{20.030 - 18.914}(2\% - 1.5\%) = 1.80\%$$

or, on an annual basis,

$$i \approx (1.0180)^{12} - 1 = 23.87\%$$

which is rather different from the advertised rate.

Example 6.9 A house is being advertised for sale by the owner. An investor estimates that the property could be rented out for $600 per month. Taxes and minor maintenance expenses are estimated at $1200 per year. The house has been recently remodeled and the tenant should have to pay all utilities. The investor thinks he could sell the house for $85 000 after 5 years. What is the largest amount that the investor can offer for the property if his MARR is 12%, compounded monthly?

The equivalent value at year 0 of the expected receipts and disbursements is given by

$$P = \$600(P/A, i_m\%, 60) + \$85\,000(P/F, i_y\%, 5) - \$1200(P/A, i_y\%, 5)$$

where $i_m = 1\% \equiv$ effective monthly MARR
$i_y = (1.01)^{12} - 1 = 12.68\% \equiv$ effective yearly MARR

Now, $(P/A, 1\%, 60) = 44.955$ (from Appendix A), and

$$(P/F, 12.68\%, 5) = (1.1268)^{-5} = 0.5505$$

$$(P/A, 12.68\%, 5) = \frac{1 - (P/F, 12.68\%, 5)}{0.1268} = 3.5449$$

Therefore,

$$P = \$600(44.955) + \$85\,000(0.5505) - \$1200(3.5449) = \$69\,511.62$$

The investor must buy the house for $69 511.62 (or less) to get an effective rate of 1% per month (or more) on his investment.

Solved Problems

6.1 Is the receipt of $4000 annually for 10 years equivalent to the receipt of $5000 annually for 8 years, if the interest rate is 8% per year, compounded annually?

$$P_1 = \$4000(P/A, 8\%, 10) = \$4000(0.14903)^{-1} = \$26\,840.23$$

$$P_2 = \$5000(P/A, 8\%, 8) = \$5000(0.17401)^{-1} = \$28\,733.98$$

The flows are not equivalent; the receipt of $5000 for 8 years gives the larger present value.

6.2 If the interest rate is 8% per year, compounded annually, what is the equivalent present value of $10 000 (*a*) 1 year from today? (*b*) 5 years from today?

(*a*) $$P = \$10\,000(P/F, 8\%, 1) = \$10\,000(1.0800)^{-1} = \$9259.26$$

(*b*) $$P = \$10\,000(P/F, 85\%, 5) = \$10\,000(1.4693)^{-1} = \$6805.96$$

6.3 What is the equivalent future value of $1000 annually for the next 9 years, if the interest rate is 8% per year, compounded annually?

$$F = \$1000(F/A, 8\%, 9) = \$1000(12.4876) = \$12\,487.60$$

6.4 A man can invest $150 000 for 12 years in a business venture and expect to receive $6000 per year in return. If his MARR is 7% per year, compounded annually, would this investment be satisfactory?

The yearly return A necessary to achieve this MARR is given by

$$\$150\,000 = A(P/A, 7\%, 12) \quad \text{or} \quad A = \$150\,000(0.12590) = \$18\,885.00 > \$6000$$

The investment would not be satisfactory.

6.5 What amount of money is equivalent to receiving \$8000 three years from today, if the interest rate is 8% per year, compounded semiannually?

$$P = \frac{F}{\left(1+\frac{r}{m}\right)^{mn}} = \frac{\$8000}{\left(1+\frac{0.08}{2}\right)^{(2)(3)}} = \$6322.52$$

6.6 A bank pays 6% interest per year, compounded (*a*) quarterly, (*b*) annually. A \$5000 deposit will grow to what amount if left in that bank for 2 years?

(*a*) $$F = \$5000(1.015)^8 = \$5632.46$$

(*b*) $$F = \$5000(1.06)^2 = \$5618.00$$

6.7 What is the equivalent present value of the following series of payments: \$5000 the first year, \$5500 the second year, and \$6000 the third year? The interest rate is 8%, compounded annually.

$$\begin{aligned} P &= [\$5000 + \$500(A/G, 8\%, 3)]\,(P/A, 8\%, 3) \\ &= [\$5000 + \$500(0.9487)]\,(0.38803)^{-1} = \$14\,108.06 \end{aligned}$$

6.8 What single amount at the end of the fourth year is equivalent to a uniform annual series of \$3000 per year for 10 years, if the interest rate is 10% per year, compounded annually?

Find the present value of the series and then move it ahead 4 years:

$$F = \$3000(P/A, 10\%, 10)\,(F/P, 10\%, 4) = \$3000(0.16275)^{-1}(1.4641) = \$26\,988.02$$

6.9 A series of 10 annual payments of \$2000 is equivalent to two equal payments, one at the end of 15 years and the other at the end of 20 years. The interest rate is 8%, compounded annually. What is the amount of the two equal payments?

$$\$2000(P/A, 8\%, 10) = X\,[(P/F, 8\%, 15) + (P/F, 8\%, 20)]$$

$$\$2000(0.14903)^{-1} = X\,[(3.1722)^{-1} + (4.6610)^{-1}]$$

$$X = \frac{\$13\,420.12}{0.31524 + 0.21455} = \$25\,331.08$$

6.10 A new corporate bond is being offered in the market for \$930. The bond has a face value of \$1000 and matures in 10 years. The issuing corporation promises to pay \$70 in interest every year. (*a*) Should an investor requiring an 8% return on investment buy this bond? (*b*) What is the company's cost of the capital raised through this bond issue if the stockbroker's fee is \$15 per bond sold?

(*a*) At the investor's rate, the present value of the bond is

$$\begin{aligned} P &= \$70(P/A, 8\%, 10) + \$1000(P/F, 8\%, 10) \\ &= \$70(6.710) + \$1000(2.1589)^{-1} = \$932.89 > \$930 \end{aligned}$$

The investor should buy the bond.

(*b*) $$\$915 = \$70(P/A, i\%, 10) + \$1000(P/F, i\%, 10)$$

for i = 8% R.H.S. = \$932.89

for i = 9% R.H.S. = \$70(6.418) + \$1000(0.4224) = \$871.66

Interpolating, $$i \approx 8\% + \frac{932.89 - 915}{932.89 - 871.66}(9\% - 8\%) = 8.29\%$$

The company's cost of capital from this bond issue is about 8.29%, compounded yearly.

6.11 Is it expedient for Thomas, who requires an 8% return on his investments, to buy an artificial Christmas tree? The tree costs $45 and is expected to last eight years. The alternative is to keep purchasing natural trees, which now cost $8 (Christmas of year 0) and are expected to increase $1 in price each coming year.

The present value at 8% of a stream of payments of $9 one year from now, $10 in two years, etc., up to $16 in eight years, is:

$$P = [\$9 + \$1(A/G, 8\%, 8)]\,(P/A, 8\%, 8) = [\$9 + \$1(3.099)]\,(5.747) = \$69.53$$

This amount, plus this year's natural tree cost ($8), greatly exceeds the artificial tree's price; Thomas should buy the artificial tree.

6.12 A mine is for sale. A mining engineer estimates that, at current production levels, the mine will yield an annual net income of $80 000 for 15 years, after which the mineral will be exhausted. If an investor's MARR is 15%, what is the maximum amount he can bid on this property?

$$P = \$80\,000(P/A, 15\%, 15) = \$80\,000(5.847) = \$467\,760$$

6.13 How long must a temporary warehouse last to be a desirable investment if it costs $16 000 to build, has annual maintenance and operating costs of $360, provides storage space valued at $3600 per year, and if the company MARR is 10%?

The yearly net income is $\$3600 - \$360 = \$3240$, and so the warehouse must last at least n years, where

$$(A/P, 10\%, n) = \frac{\$3240}{\$16\,000} = 0.2025$$

Solving by Table 3-1, or by scanning the table of A/P in Appendix A, we find that $n = 8$ years.

6.14 For what value of X are the following two cash flows equivalent at a 10% interest rate?

End of Year	0	1	2	3	4
Flow A, $	−10 000	5000	5000	5000	5000
Flow B, $	X	3500	4500	5500	6500

Equating

$$P_A = -\$10\,000 + \$5000(P/A, 10\%, 4)$$

and

$$P_B = X + [\$3500 + \$1000(A/G, 10\%, 4)]\,(P/A, 10\%, 4)$$

we obtain

$$\begin{aligned} X &= -\$10\,000 + [\$1500 - \$1000(A/G, 10\%, 4)]\,(P/A, 10\%, 4) \\ &= -\$10\,000 + [\$1500 - \$1000(1.3812)]\,(0.31547)^{-1} \\ &= -\$9623 \end{aligned}$$

Supplementary Problems

6.15 What series of equal annual payments is economically equivalent to the investment of a present amount of $5000 for 5 years at 12%, compounded annually? *Ans.* $1387.03

6.16 What single amount at the end of the fifth year is equivalent to a uniform annual series of $2000 per year for 10 years, if the interest rate is 10%, compounded annually? *Ans.* $19 791.09

6.17 A series of 12 annual payments of $2000 is equivalent to three equal payments, one each at the end of 12 years, 15 years, and 20 years. The interest rate is 10%, compound annually. What is the amount of the three equal payments? *Ans.* $19 284.55

6.18 What is the equivalent present value of the following series of payments: $7000 the first year, $6500 the second year, $6000 the third year, $5500 the fourth year, and $5000 the fifth year? The interest rate is 10%, compounded annually. *Ans.* $23 104.44

6.19 Is a series of 100 equal quarterly payments of $800 equivalent to a present amount of $35 000 if the interest rate is 8% per year, compounded quarterly?
Ans. No; it is equivalent to exactly $34 482.76.

6.20 Machine X will produce cost savings of $5000 per year for four years; machine Y will produce cost savings of $4000 per year for five years. If the interest rate is 10%, compound annually, are these two machines economically equivalent in terms of the present value of their cost savings?
Ans. No: $P_X = \$15\,849.37$ and $P_Y = \$15\,163.00$.

6.21 Find the uniform annual series of seven payments that would be equivalent to the following gradient series: $500 initially, with a $50 increment per year, for a total of seven years. The interest rate is 12%, compound annually. *Ans.* $627.58

6.22 A woman can invest $100 000 for 15 years in a bank and expect to receive a yearly return of $10 000. The woman's objective is to earn 12% per year, compounded annually, on her investments. Is this objective met by the bank plan? *Ans.* No; the woman requires a return of $14 682 per year.

6.23 Machine A will save $5000 per year for 6 years; machine B will save $6000 per year for 5 years. If the interest rate is 10%, compounded annually, do these two machines have equal future values at the end of the sixth year? *Ans.* No: $F_A = \$38\,578$ and $F_B = \$36\,630.60 \times 1.10 = \$40\,293.66$.

6.24 The XYZ Bank advertises it will pay $3869.70 in cash at the end of 20 years to anyone who deposits $1000; the ABC Bank states that it pays 10% per year, compounded annually, on all deposits left one year or more. Which bank is paying the higher interest rate, and by how much?
Ans. ABC; XYZ is paying only 7% per year, compounded annually.

6.25 A series of quarterly payments of $1000 for 25 years is economically equivalent to what present sum, if the quarterly payments are invested at an annual rate of 8%, compounded quarterly?
Ans. $43 103.45

6.26 What series of equal annual payments is economically equivalent to the investment of a present amount of $5000 for 5 years at 12%, compounded annually? *Ans.* $1387.03

6.27 A promissory note has outstanding payments of $650 at the end of each of the next five years. What market price would be paid for this note by an investor who requires a 12% yield on his investments, compounded quarterly? *Ans.* $2311.47

6.28 A loan of $750 is to be repaid in 18 equal monthly installments, computed as follows:

Loan	$750
Interest at 1% per month for 18 months	144
Credit check and processing fee	60
	$954

Monthly payment: $954/18 = $53

What (*a*) nominal and (*b*) effective annual interest rates are being charged?
Ans. (*a*) 32.15%; (*b*) 37.34%

Chapter 7

PW, FW, EUAS/EUAC

This chapter treats several valuation methods which are useful in deciding among economic alternatives. Its sequel, Chapter 8, is devoted to techniques that are primarily used for analyzing proposed capital investments.

7.1 PRESENT WORTH

The *present worth* (PW) or *present value* (PV) of a given series of cash flows is the equivalent value of the cash flows at the end of year 0 (i.e., at the beginning of year 1). For the case of annual compounding,

$$\begin{aligned} PW &= CF_0 + CF_1\,(P/F, i\%, 1) + CF_2\,(P/F, i\%, 2) + \cdots + CF_n\,(P/F, i\%, n) \\ &\equiv \sum_{j=0}^{n} CF_j\,(P/F, i\%, j) \end{aligned} \tag{7.1}$$

Here, CF_j is the (positive or negative) cash flow for the jth year, $(P/F, i\%, j) = (1+i)^{-j}$, and n is the total number of years.

Example 7.1 Determine the present worth of the following series of cash flows, based on an interest rate of 12% per year, compounded annually: $0 (end of year 0), $1000 (1), $2000 (2), $3000 (3), $4000 (4), $4000 (5), $4000 (6).

By (*7.1*),

$$\begin{aligned} PW &= \$1000(P/F, 12\%, 1) + \$2000(P/F, 12\%, 2) + \$3000(P/F, 12\%, 3) \\ &\quad + \$4000[(P/F, 12\%, 4) + (P/F, 12\%, 5) + (P/F, 12\%, 6)] \\ &= \$1000(0.8929) + \$2000(0.7972) + \$3000(0.7118) + \$4000(1.7096) = \$11\,460.70 \end{aligned}$$

Alternatively, since the cash flows compose a gradient series followed by a uniform series,

$$\begin{aligned} PW &= [\$1000 + \$1000(A/G, 12\%, 4)]\,(P/A, 12\%, 4) + \$4000(P/A, 12\%, 2)\,(P/F, 12\%, 4) \\ &= [\$1000 + \$1000(1.3589)](3.0374) + \$4000(1.6901)(0.6355) \\ &= \$7164.92 + \$4296.23 = \$11\,461.15 \end{aligned}$$

Figure 7-1 diagrams this latter solution (which agrees with the former up to roundoff errors).

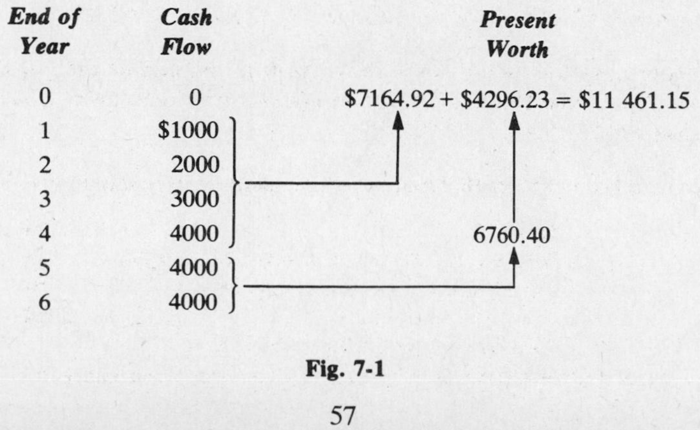

End of Year	*Cash Flow*	*Present Worth*
0	0	$7164.92 + $4296.23 = $11 461.15
1	$1000	
2	2000	
3	3000	
4	4000	6760.40
5	4000	
6	4000	

Fig. 7-1

Equation (*7.1*) is often applied to problems involving an initial cash outflow followed by a series of cash inflows; i.e., $CF_0 < 0$ and $CF_j > 0 \quad (j > 0)$. In such cases, the PW is renamed the *net present worth* (NPW) or the *net present value* (NPV). Clearly the NPW is a monotonically decreasing function of the interest rate i (because, as i increases, the positive flows—and only these—are increasingly discounted). Applications of the NPW will be given in Chapter 8.

7.2 FUTURE WORTH

Given a series of cash flows as in Section 7.1, the *future worth* (FW) of the series is its equivalent value at the end of year n. Assuming annual compounding,

$$FW = CF_0\,(F/P, i\%, n) + CF_1\,(F/P, i\%, n-1) + CF_2\,(F/P, i\%, n-2) + \cdots + CF_{n-1}\,(F/P, i\%, 1) + CF_n$$
$$\equiv \sum_{j=0}^{n} CF_j\,(F/P, i\%, n-j) \tag{7.2}$$

From the theory of Chapter 2, the future worth is related to the present worth via

$$FW = PW \times (F/P, i\%, n) \tag{7.3}$$

Example 7.2 Determine the future worth of the cash flows given in Example 7.1, based on an interest rate of 12% per year, compounded annually, using (*a*) (*7.2*), (*b*) (*7.3*).

(*a*)
$$\begin{aligned} FW &= \$1000(F/P, 12\%, 5) + \$2000(F/P, 12\%, 4) + \$3000(F/P, 12\%, 3) \\ &\quad + \$4000[(F/P, 12\%, 2) + (F/P, 12\%, 1) + 1] \\ &= \$1000(1.7623) + \$2000(1.5735) + \$3000(1.4049) + \$4000(3.3744) = \$22\,621.60 \end{aligned}$$

(*b*)
$$FW = \$11\,460.70\,(F/P, 12\%, 6) = \$11\,460.70\,(1.9738) = \$22\,621.13$$

The variation of FW with i, in the special case $CF_0 < 0$ and $CF_j > 0 \quad (j > 0)$, is a little more complicated than the analogous variation of PW. If $|CF_0|$ is very small in comparison to the positive flows, then the future value of the positive flows—and along with it the FW—will *increase* as i increases. However, if $|CF_0|$ is sufficiently large (as it will be, in practical applications), FW will be a monotonically decreasing function of i, like PW.

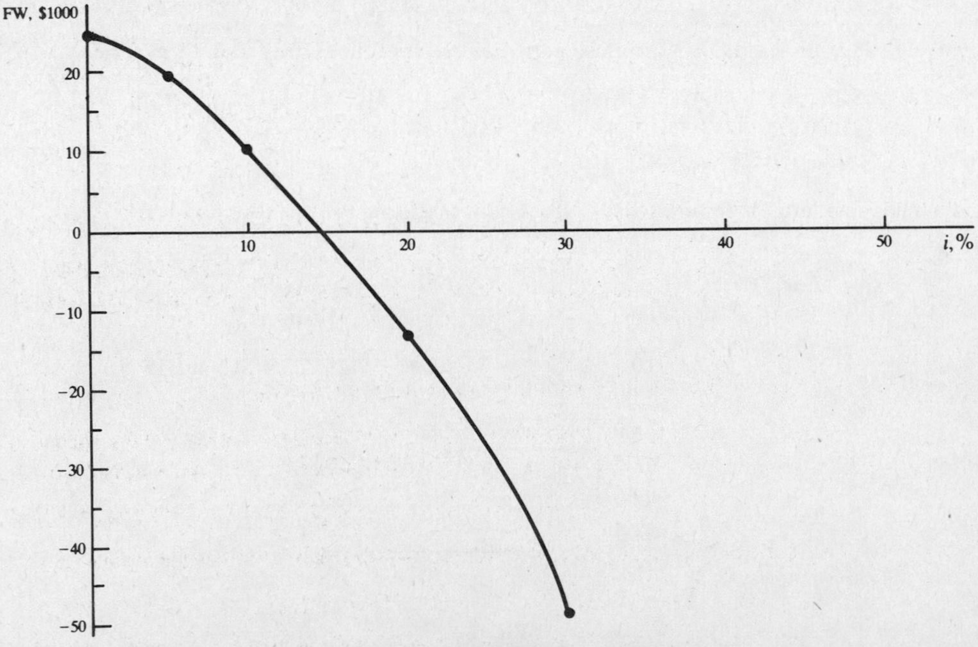

Fig. 7-2

Example 7.3 Given the series

End of Year	0	1	2	3	4	5
Cash Flow, $1000	−50	15	15	15	15	15

determine the future worth of the series at annual interest rates 0%, 5%, 10%, 20%, 30%, and 50%. Graph your results.

Using the equation

$$\text{FW} = -\$50\,000(F/P, i\%, 5) + \$15\,000(F/A, i\%, 5)$$

instead of (*7.2*), we calculate the following points (i, FW): (0%, $25 000), (5%, $19 070), (10%, $11 051), (20%, −$12 792), (30%, −$49 998), (50%, −$181 870). These points are plotted to give the curve of Fig. 7-2. It is seen that FW rapidly decreases with i, becoming zero at $i \approx 15\%$ (more precisely, at $i = 15.26\%$). In view of (*7.3*), PW must vanish at this same interest rate.

7.3 EQUIVALENT UNIFORM ANNUAL SERIES

The *equivalent uniform annual series* (EUAS) is obtained by converting the equivalent value (at a specified time, usually the present) of a given set of cash flows into a series of uniform annual payments. Thus, if interest is compounded annually, we can write

$$\text{EUAS} = \text{PW} \times (A/P, i\%, n) \tag{7.4}$$

or, substituting (*7.1*),

$$\text{EUAS} = \left[\sum_{j=0}^{n} \text{CF}_j \,(P/F, i\%, j)\right] \times (A/P, i\%, n) \tag{7.5}$$

The EUAS is widely used for analyzing decision alternatives.

Example 7.4 The EUAS is particularly convenient when studying a *repeated cycle* of cash flows, as exemplified in Fig. 7-3. Applying (*7.5*) to the first 3-year cycle, we have, assuming $i = 9\%$:

$$\begin{aligned}\text{EUAS} &= [-\$600 + \$400(P/F, 9\%, 1) + \$300(P/F, 9\%, 2) + \$500(P/F, 9\%, 3)]\,(A/P, 9\%, 3)\\ &= [-\$600 + \$400(0.9174) + \$300(0.8417) + \$500(0.7722)](0.3951)\\ &= \$160.24 \text{ per year for 3 years}\end{aligned}$$

End of Year	***Disbursements (Cash Outflows)***	***Receipts (Cash Inflows)***
0	$600	
1		$400
2		300
3	600	500
4		400
5		300
6	600	500
7		400
8		300
9	600	500
10		400
11		300
12		500

Fig. 7-3

Because each cycle has this same EUAS (relative to its starting year), the EUAS for the entire series is \$160.24 per year for 12 years.

In the important special case $CF_0 < 0$ and $CF_j > 0$ $(j > 0)$, it can be shown that the EUAS decreases, in almost linear fashion, as i increases. Furthermore, as is shown by (*7.4*), it becomes zero at the same value of i for which PW (and FW) becomes zero.

If many or all of the cash flows are negative (i.e., are *costs*), it may be convenient to deal with the negative of the EUAS; we call this quantity the *equivalent uniform annual cost* (EUAC). It is clear that, in calculating the EUAS or EUAC, we may neglect any *constant* yearly cash flow (e.g., a fixed annual maintenance charge), and simply add in that constant amount at the end.

7.4 CAPITAL RECOVERY

Let us apply the notion of EUAS/EUAC to an asset whose series of cash flows consists of just two terms: an original cost, P, and an (actual or estimated) *salvage value*, SV, at the end of n years. For this series,

$$\text{PW} = -P + \text{SV}\,(P/F, i\%, n)$$

and (*7.4*) gives

$$\begin{aligned}\text{EUAC} = -\text{EUAS} &= [P - \text{SV}\,(P/F, i\%, n)]\,(A/P, i\%, n)\\ &= P\,(A/P, i\%, n) - \text{SV}\,(A/F, i\%, n)\\ &= (P - \text{SV})(A/P, i\%, n) + i\,\text{SV}\end{aligned} \tag{7.6}$$

in which (*3.5*) was used in the last step. Because here the EUAC represents the difference between the annualized cost of the asset and the annualized salvage value, it is retitled the *capital recovery* (*cost*) and resymbolized CR.

Example 7.5 A machine which costs \$50 000 when new has a 10-year lifetime and a salvage value equal to 10% of its original value. Determine the capital recovery, based upon an interest rate of 8% per year, compounded annually.

From (*7.6*),

$$\begin{aligned}\text{CR} &= (\$50\,000 - \$5000)(A/P, 8\%, 10) + 0.08(\$5000)\\ &= \$45\,000(0.1490) + \$5000(0.08) = \$7105 \text{ per year}\end{aligned}$$

This value represents the annualized net cost of the machine.

Unlike the EUAS/EUAC in general, the CR does *not* take into account operating or maintenance expenses associated with the asset.

7.5 CAPITALIZED EQUIVALENT

Suppose that a given sum of money, P, earns interest at an annual rate i. If the interest is withdrawn at the end of each year but the principal is left intact, then a perpetual series of uniform annual payments will be obtained, the amount of each payment being

$$A = iP \tag{7.7}$$

[(*7.7*) also follows from (*3.6*).]

Looking at matters the other way round, we call P the *capitalized equivalent* (CE) of the perpetual annual payments A, and write

$$\text{CE} = \frac{A}{i} \tag{7.8}$$

Solved Problems

7.1 Determine the present worth of the following cash flows, based on an interest rate of (*a*) 10% per year, (*b*) 15% per year, compounded annually. Explain the results.

End of Year	0	1	2	3	4	5
Cash Flow, $1000	3	6	4	1	7	5

(*a*)
$$\begin{aligned} PW &= \$3000 + \$6000(P/F, 10\%, 1) + \$4000(P/F, 10\%, 2) \\ &\quad + \$1000(P/F, 10\%, 3) + \$7000(P/F, 10\%, 4) + \$5000(P/F, 10\%, 5) \\ &= \$3000 + \$6000(0.9091) + \$4000(0.8264) \\ &\quad + \$1000(0.7513) + \$7000(0.6830) + \$5000(0.6209) \\ &= \$20\,397.00 \end{aligned}$$

(*b*)
$$\begin{aligned} PW &= \$3000 + \$6000(0.8696) + \$4000(0.7561) \\ &\quad + \$1000(0.6575) + \$7000(0.5718) + \$5000(0.4972) \\ &= \$18\,388.10 \end{aligned}$$

The PW at $i = 15\%$ is smaller than the PW at $i = 10\%$; at the higher interest rate, a smaller present sum can generate the given series of *positive* payments.

7.2 Compute the corresponding future worths of the cash flows in Problem 7.1. Explain the results.

Since the present worths are known, it is simplest to use (*7.3*).

(*a*)
$$FW = (\$20\,397.00)(F/P, 10\%, 5) = (\$20\,397.00)(1.6105) = \$32\,849.37$$

(*b*)
$$FW = (\$18\,388.10)(F/P, 15\%, 5) = (\$18\,388.10)(2.0114) = \$36\,985.82$$

At the higher interest rate, the *positive* payments accumulate to a higher future value.

7.3 Compute the present worth of the following cash flows at (*a*) $i = 6\%$ per year, (*b*) $i = 15\%$ per year, compounded annually. Explain the results.

End of Year	0	1	2	3	4
Cash Flow, $1000	−40	12	12	12	12

(*a*)
$$PW = -\$40\,000 + \$12\,000(P/A, 6\%, 4) = -\$40\,000 + \$12\,000(0.28859)^{-1} = +\$1581.48$$

(*b*)
$$PW = -\$40\,000 + \$12\,000(P/A, 15\%, 4) = -\$40\,000 + \$12\,000(0.35027)^{-1} = -\$5740.71$$

In accordance with the discussion in Section 7.1, the PW declines as the interest rate increases.

7.4 Compute the future worth and the equivalent uniform annual series value for the cash flows in Problem 7.3. Explain the results.

Use (*7.3*) and (*7.4*).

(*a*)
$$FW = (\$1581.38)(F/P, 6\%, 4) = (\$1581.48)(1.2625) = +\$1996.62$$
$$EUAS = (\$1581.48)(A/P, 6\%, 4) = (\$1581.48)(0.28859) = +\$456.40$$

(*b*)

$$FW = (-\$5740.71)(F/P, 15\%, 4) = (-\$5740.71)(1.7490) = -\$10\,040.50$$
$$EUAS = (-\$5740.71)(A/P, 15\%, 4) = (-\$5740.71)(0.35027) = -\$2010.80$$

The behavior discussed in Sections 7.2 and 7.3 is exhibited here: both FW and EUAS decrease as i increases.

7.5 An engineer is thinking of starting a part-time consulting business next September 5, on his 40th birthday. He expects the business will require an initial cash outlay of \$5000, to come from his savings, and will cost \$500 per year to operate; the business ought to generate \$2000 per year in cash receipts. During the 20 years that he expects to operate the business, he plans to deposit the annual net proceeds in a bank each year, at an interest rate of 8% per year, compounded annually. When he retires, on his 60th birthday, the engineer expects to invest whatever proceeds plus interest he then has from the business in a long-term savings plan that pays 10% per year, compounded annually. What is the maximum amount he could withdraw from the savings plan each year during his retirement and still have the funds last 15 years?

The net proceeds from the business will be \$2000 − \$500 = \$1500 per year. Therefore, at the end of 20 years, the engineer will have

$$\begin{aligned} FW &= \$1500(F/A, 8\%, 20) - \$5000(F/P, 8\%, 20) \\ &= \$1500(45.7620) - \$5000(4.6610) = \$45\,338 \end{aligned}$$

The maximum annual amount he could withdraw is therefore

$$A = \$45\,338(A/P, 10\%, 15) = \$45\,338(0.13147) = \$5960.59$$

7.6 Let $i = 15\%$ per year, compounded annually. Determine the present worth of the following cash flows:

End of Year	0	1	2	3	4
Cash Flow, \$1000	−10	2	2	6	6

Analyze the last two flows as \$2000 + \$4000. Then,

$$\begin{aligned} PW &= -\$10\,000 + \$2000(P/A, 15\%, 4) + \$4000(P/A, 15\%, 2)(P/F, 15\%, 2) \\ &= -\$10\,000 + \$2000(0.35027)^{-1} + \$4000(0.61512)^{-1}(1.3225)^{-1} = \$626.93 \end{aligned}$$

7.7 Determine the EUAS for the repeated cycle of disbursements and receipts shown in Fig. 7-4, if the interest rate is 10% per year, compounded annually.

$$\begin{aligned} EUAS &= [-\$500 + \$200(P/F, 10\%, 1) + \$150(P/F, 10\%, 2) \\ &\quad + \$300(P/F, 10\%, 3) + \$400(P/F, 10\%, 4)]\,(A/P, 10\%, 4) \\ &= [-\$500 + \$200(1.1000)^{-1} + \$150(1.2100)^{-1} + \$300(1.3310)^{-1} + \$400(1.4641)^{-1}](0.31547) \\ &= \$96.03 \text{ per year for 16 years} \end{aligned}$$

7.8 A machine costs \$40 000 to purchase and \$10 000 per year to operate. The machine has no salvage value, and a 10-year life. If $i = 10\%$ per year, compounded annually, what is the equivalent uniform annual cost of the machine?

The EUAC is given by (*7.4*) or (*7.5*), with costs counted as positive, together with the fixed operating cost.

$$EUAC = \$40\,000(A/P, 10\%, 10) + \$10\,000 = \$40\,000(0.16275) + \$10\,000 = \$16\,510$$

End of Year	Disbursements	Receipts
0	$500	
1		$200
2		150
3		300
4	500	400
5		200
6		150
7		300
8	500	400
9		200
10		150
11		300
12	500	400
13		200
14		150
15		300
16		400

Fig. 7-4

7.9 A new bridge with a 100-year life is expected to have an initial cost of $20 million. This bridge must be resurfaced every five years, at a cost of $1 million. The annual inspection and operating costs are estimated to be $50 000. Determine the present-worth cost of the bridge using the capitalized equivalent approach (i.e., take the life of the bridge as infinite). The interest rate is 10% per year, compounded annually.

The present worth of the nonrecurring cost is simply $P_1 = \$20$ million. The recurring $1 million cost is equivalent to

$$A_1 = (\$1\,000\,000)(A/F, 10\%, 5) = (\$1\,000\,000)(6.1051)^{-1} = \$163\,797 \text{ per year}$$

Thus, there are two annual costs, $A_1 = \$163\,797$ and $A_2 = \$50\,000$; their combined capitalized equivalent is

$$\text{CE} = \frac{\$163\,797 + \$50\,000}{0.10} = \$2\,137\,970$$

and the total present-worth cost is

$$P_1 + \text{CE} = \$20\,000\,000 + \$2\,137\,970 = \$22\,137\,970$$

7.10 Determine the approximate size of the annual payment needed to retire $70 000 000 in bonds issued by a city to build a dam. The bonds must be repaid over a 50-year period, and they earn interest at an annual rate of 6%, compounded annually.

To the extent that 50 years may be taken as infinite,

$$A = \text{CE} \times i = (\$70\,000\,000)(0.06) = \$4\,200\,000$$

7.11 A machine that cost $30 000 new has an 8-year life and a salvage value equal to 10% of its original cost. The annual maintenance cost of this machine is $1000 the first year, with an increase of $200 each year hereafter; the annual operating cost is $800 per year. Determine the EUAC of this machine if the interest rate is 10% per year, compounded annually.

The EUAC is the sum of three terms: the CR given by (*7.6*); the EUAC for the \$200 gradient; and the fixed annual cost, \$1000 + \$800 = \$1800.

$$\text{EUAC} = (\$30\,000 - \$3000)(A/P, 10\%, 8) + (0.10)(\$3000) + \$200(A/G, 10\%, 8) + \$1800$$
$$= \$27\,000(0.18744) + \$300 + \$200(3.0045) + \$1800 = \$7761.78$$

Supplementary Problems

7.12 Find the present worth of the machine of Problem 7.8.
Ans. PW = −\$101 443.93 (the negative value indicates a cash outflow or cost)

7.13 Rework Problem 7.8, using annual interest rates of (*a*) 5%, (*b*) 15%, and (*c*) 20%, compounded annually. (*d*) Comment on the results.
Ans. (*a*) EUAC = \$15 180; (*b*) EUAC = \$17 790; (*c*) EUAC = \$19 541; (*d*) the EUAC increases with i, since all costs are positive and since the value of $(A/P, i\%, 10)$ increases as i increases.

7.14 A machine costs \$30 000 to purchase and \$12 000 per year to operate. The machine has a 10-year life and no salvage value. Determine the EUAC of this machine at annual interest rates of (*a*) 5%, (*b*) 15%, and (*c*) 20%, compounded annually. *Ans.* (*a*) \$15 885.00; (*b*) \$17 977.50; (*c*) \$19 155.60

7.15 A used machine costs \$20 000 to purchase. It has an annual maintenance cost of \$20 000, a salvage value of \$5000, and a 10-year life. If the interest rate is 10% per year, compounded annually, what is the present-worth *cost* of the machine? *Ans.* \$140 960.12

7.16 Rework Problem 7.6 for $i = 8\%$ per year, compounded annually. *Ans.* \$2739.71

7.17 Use (*7.2*) to determine the future worth of the cash flows in Problem 7.6, for $i = 8\%$ per year, compounded annually. *Ans.* \$3727.20

7.18 Calculate by (*7.2*) the future worth of the cash flows of Problem 7.6, using (*a*) $i = 15\%$ per year, (*b*) $i = 8\%$ per year, compounded annually. *Ans.* (*a*) \$1096.50; (*b*) \$3727.38

7.19 Compute the future-worth cost of the machine of Problem 7.15. *Ans.* \$365 608.25

7.20 Compute the EUAC of the machine of Problem 7.15, using the result of Problem 7.15.
Ans. \$22 941.26 per year

7.21 Compute the capital recovery for the machine of Problem 7.15, using (*7.6*). Compare with the answer to Problem 7.20. *Ans.* \$2941.25 = \$22 941.26 − \$20 000

7.22 Determine the present worth of the following series of cash flows, given $i = 15\%$: $CF_0 = -\$10\,000$, $CF_1 = CF_2 = CF_3 = CF_4 = \$5000$, $CF_5 = -\$2000$, $CF_6 = \$3000$. *Ans.* \$6566.15

7.23 Calculate the present worth of the following cash flows, when $i = 10\%$ per year, compounded annually.

End of Year	0	1	2	3	4
Cash Flow, \$1000	−10	4	4	4	4

Ans. \$2679.49

7.24 Calculate the present worth of the following cash flows, if the interest rate is 12% per year, compounded annually.

End of Year	0	1	2	3	4
Cash Flow, $1000	−15	5	6	2	4

Ans. −$1786.86

7.25 A machine which costs $100 000 when new has a lifetime of 15 years and a salvage value equal to 20% of its original cost. Determine the capital recovery for this machine, if the interest rate is 10% per year, compounded annually. *Ans.* $12 517.60 per year

7.26 Repeat Problem 7.25, using a salvage value equal to 5% of the machine's original cost. *Ans.* $12 989.56 per year

7.27 Determine the capital recovery for the machine in Problem 7.25, if the interest rate is 15% per year, compounded annually, and total operating and maintenance costs are $2000 per year. *Ans.* $16 681.60 per year

7.28 Determine the amount of money required to generate an infinite number of annual payments of $5000 each, if the interest rate is 10% per year, (*a*) compounded annually, (*b*) compounded continuously. *Ans.* (*a*) $50 000; (*b*) $47 528.52

7.29 Mr. Diamond expects to invest $1000 per year for each of the next 20 years in an investment plan that pays 10% per year, compounded annually. At the end of the 20th year, he expects to withdraw the balance in his investment plan and deposit it in a savings account. This savings account pays 6% per year, compounded *monthly*. Mr. Diamond wants to withdraw a fixed amount from this savings account each month, for a total of five years. How large may this fixed amount be? *Ans.* $1107.13

7.30 Repeat Problem 7.7 for an interest rate of 15% per year, compounded annually. *Ans.* $74.71 per year for 16 years

7.31 Costs of $10 000, $20 000, and $23 000 are incurred at the ends of three successive years. Find the EUAC for k repetitions of the cycle, if (*a*) $i = 10\%$ per year, (*b*) $i = 15\%$ per year, compounded annually. *Ans.* (*a*) $17 250.55 per year for $3k$ years; (*b*) $21 322.10 per year for $3k$ years

7.32 A flood-control dam with a 100-year life has an initial cost of $15 million. The gates in the dam must be replaced every five years, at a cost of $2 million. If the interest rate is 8% per year, compounded annually, what is the capitalized equivalent of the annual cost of the dam? *Ans.* $4 261 412

7.33 What is the total present worth of the dam described in Problem 7.32? *Ans.* $19 261 412

7.34 Assuming that money earns 10% a year, which would be the better arrangement for leasing an electron microscope: (1) paying a deposit of $100 000, to be returned at the end of the lease period; or (2) paying $10 000 a year for as long as the device is kept?
Ans. For (1), (*7.6*) gives EUAC = iP = (0.10)($100 000) = $10 000, and so the two plans are equivalent.

Chapter 8

Net Present Value, Rate of Return, Payback Period, Benefit-Cost Ratio

This chapter continues the ideas developed in Chapter 7, particularly as they are applied in deciding among alternative capital investments.

8.1 NET PRESENT VALUE

The definition, (*7.1*), of the NPV is repeated here:

$$\text{NPV} = -|\text{CF}_0| + \sum_{j=1}^{n} \text{CF}_j\,(P/F, i\%, j) \tag{8.1}$$

in which the notation emphasizes our assumption that the initial cash flow, CF_0, is negative (a capital outlay). No assumption is made concerning the signs of the remaining CF_j, although often these terms will all be positive (revenues). In the special case $\text{CF}_j = A \quad (j = 1, 2, \ldots, n)$, (*8.1*) becomes, in view of (*3.4*),

$$\text{NPV} = -|\text{CF}_0| + A\,(P/A, i\%, n) \tag{8.2}$$

as it must. Another name for the NPV is the *discounted cash flow*, or DCF.

From (*8.1*) it is seen that the NPV is positive when and only when the total value of the returns CF_j (in year 0 dollars) exceeds the amount invested, $|\text{CF}_0|$ (year 0 dollars); that is to say, when and only when the original amount, earning compound interest at rate i for n years, would be insufficient to generate the returns. For a proposed investment to be economically acceptable, the NPV must be positive or, at worst, zero (in which case the investment of $|\text{CF}_0|$ would just suffice to yield the revenues CF_j).

Example 8.1 The cash flows associated with a milling machine are $\text{CF}_0 = -\$50\,000$. $\text{CF}_j = \$15\,000$ $(j = 1, \ldots, 5)$. Use (*8.2*) to determine the economic acceptability of this machine at interest rates of (*a*) 10%, (*b*) 15%, and (*c*) 20% per year, all compounded annually.

(*a*) $$\text{NPV} = -\$50\,000 + \$15\,000(P/A, 10\%, 5) = -\$50\,000 + \$15\,000(0.26380)^{-1} = \$6861.26$$

(*b*) $$\text{NPV} = -\$50\,000 + \$15\,000(P/A, 15\%, 5) = -\$50\,000 + \$15\,000(0.29832)^{-1} = \$281.58$$

(*c*) $$\text{NPV} = -\$50\,000 + \$15\,000(P/A, 20\%, 5) = -\$50\,000 + \$15\,000(0.33438)^{-1} = -\$5140.85$$

The machine is seen to be an economically acceptable investment when the interest rate is 10%, and (barely) when the interest rate is 15%. It is *not* economically justifiable to buy the machine if the interest rate is 20%.

8.2 RATE OF RETURN

The *rate of return* (ROR) for a series of cash flows is that particular value, i^*, of the interest rate for which the NPV vanishes. Thus, if we plot the NPV as a function of i, using (*8.1*) or (*8.2*), the curve will cross the i-axis at i^*. Alternatively, we could find, by trial and error, i-values for which the NPV is slightly positive and slightly negative, and interpolate linearly between them for i^*. If a more accurate approximation for i^* is required, the Newton-Raphson iteration method or another numerical technique can be used to solve (*8.1*) or (*8.2*) for i, with the left side replaced by zero.

Example 8.2 Find the ROR for the machine of Example 8.1.

By linear interpolation between the results of Example 8.1(*b*) and (*c*):

i	NPV
15%	\$281.58
i^*	0
20%	−\$5140.85

$$i^* \approx 15\% + \frac{0 - \$281.58}{-\$5140.85 - \$281.58}(20\% - 15\%) = 15.26\%$$

Case of a Single Sign-Reversal

From Section 7.1, we know that when $CF_0 < 0$ and $CF_j > 0$ $(j > 0)$—that is, when there is just one reversal of sign in the sequence $CF_0, CF_1, CF_2, \ldots, CF_n$—the NPV is a monotone decreasing function of i, and so i^* is *uniquely* determined. Moreover, at this unique ROR, the FW and EUAS are zero. (Compare Example 8.2 with Example 7.3 and Fig. 7-2.)

Case of Multiple Sign-Reversals

When the sequence $CF_0, CF_1, CF_2, \ldots, CF_n$ shows more than one reversal of sign, it is possible that NPV = 0 for several values of the interest rate; there could thus be several rates of return.

Example 8.3 For the series of cash flows

End of Year	0	1	2	3	4	5
Cash Flow, \$1000	−3	0	6	6	0	−10

determine the NPV at annual interest rates 0%, 5%, 10%, 20%, 30%, 50%, and 70%. From a graph of the results, find the rate(s) of return.

For the given flows,

$$\text{NPV} = -\$3000 + \$6000(P/A, i\%, 2)(P/F, i\%, 1) - \$10\,000(P/F, i\%, 5)$$

and evaluation at the specified interest rates gives the points

i, %	0	5	10	20	30	50	70
NPV, \$	−1000	−210	257	620	588	128	−407

which are plotted in Fig. 8-1. It is seen that there are two rates of return in this case, $i^* \approx 7\%$ and $i^* \approx 54\%$.

An upper bound on the number of (positive) rates of return may be obtained by deriving from (*8.1*) the polynomial equation

$$CF_0x^n + CF_1x^{n-1} + \cdots + CF_{n-1}x + CF_n = 0 \tag{8.3}$$

for $x \equiv 1 + i^*$. According to Descartes' rule of signs, *the number of positive real roots x cannot exceed the number of sign changes in the series of coefficients* $CF_0, CF_1, \ldots, CF_n$. Now, a positive x might correspond to a negative i^*; therefore, the number of sign changes is a fortiori an upper limit on the number of i^*-values. In particular, if there are *no* sign changes, there is *no* ROR for the given flows.

Example 8.4 There are two sign reversals in the cash flows of Example 8.3, and two values of i^* were found. In this case, the upper bound is actually attained.

If multiple i^*-values exist, it is usually better to abandon the ROR method and instead to investigate the sign of the NPV for various assumed values of the interest rate.

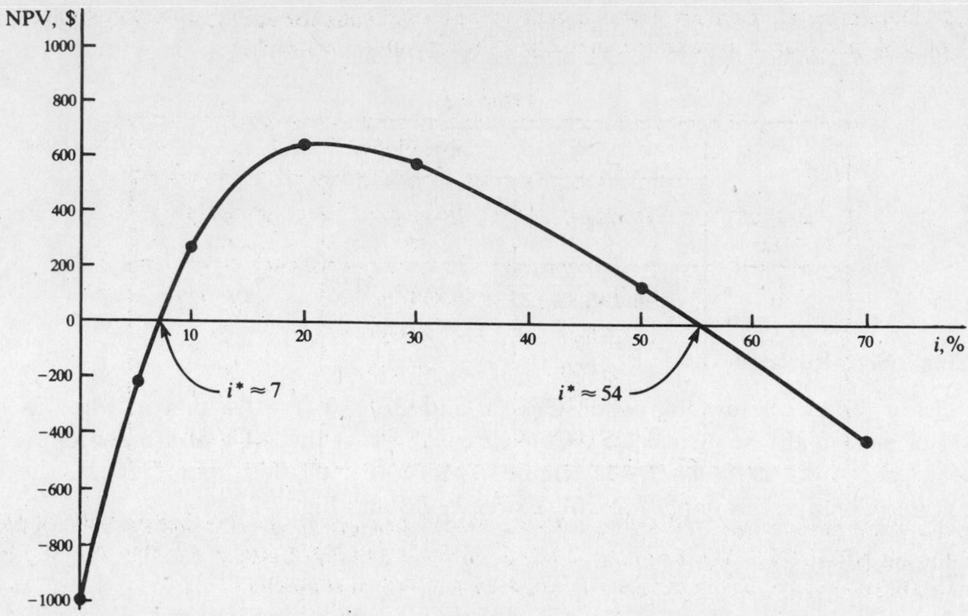

Fig. 8-1

8.3 PAYBACK PERIOD

The *payback period* (PBP) is the time required for an initial investment to be recovered, *neglecting the time value of money*. Thus, if $|CF_0|$ represents the initial investment and CF_j is the net cash inflow for the jth year ($j = 1, 2, \ldots, n$), the payback period satisfies

$$|CF_0| = \sum_{j=1}^{PBP} CF_j \tag{8.4}$$

If the yearly cash inflows are equal, or if an average value is used, then (*8.4*) simplifies to

$$PBP = \frac{|CF_0|}{YCF} \tag{8.5}$$

where YCF represents the (average) yearly cash inflow.

Example 8.5 Determine the payback period for a proposed investment as follows:

End of Year	0	1	2	3	4	5
Cash Flow, $1000	−50	10	12	15	18	20

The sum of the first three yearly cash inflows, $37 000, is less than the initial investment, $50 000; but the sum of the first four yearly cash inflows, $55 000, exceeds the initial investment. Hence the payback period will be somewhere between 3 and 4 years. Linear interpolation yields

$$PBP \approx 3 + \frac{\$50\,000 - \$37\,000}{\$55\,000 - \$37\,000}(4-3) = 3.72 \text{ years}$$

Because it ignores the time value of money, the payback method should not be used in place of the other methods discussed above. On the other hand, the payback method is valuable for a *secondary analysis*, when the NPV or ROR is used as the primary method. As will be further discussed in Chapter 9, there are many practical examples where an investment is sought with a high rate of return and a sufficiently short payback period.

Example 8.6 Determine the payback period and the net present value for each proposal in Table 8-1, using an interest rate of 10% per year, compounded annually. Which proposal is best?

Table 8-1

End of Year	Cash Flows		
	Proposal A	Proposal B	Proposal C
0	−$75 000	−$75 000	−$75 000
1	25 000	20 000	0
2	25 000	25 000	0
3	25 000	30 000	0
4	25 000	35 000	$130 000

Proposals A and B each have a 3-year payback period; however, proposal A has an NPV of $4248, while proposal B has an NPV of $10 289. Proposal C has an NPV of $13 792, but it has a 3.58-year payback period (assuming the $130 000 to be evenly spread over the fourth year). In summary:

Proposal	NPV, $	PBP, years
A	4 248	3.00
B	10 289	3.00
C	13 792	3.58

Since proposal A is inferior to proposal B, it can be eliminated from further consideration. Proposal C is economically superior to proposal B, but its longer payback period might constrain the decision maker to choose proposal B instead (e.g., if the firm were cash-poor and could not afford to wait until the end of the third year to receive any cash inflows). Thus, the choice of the best investment alternative may involve a trade-off among two or more objectives, with the PBP providing important secondary information.

8.4 BENEFIT-COST RATIO

The *benefit-cost ratio* (BCR) is often used to assess the value of a municipal project in relation to its cost; it is defined as

$$\text{BCR} \equiv \frac{B - D}{C} \tag{8.6}$$

where B represents the equivalent value of the benefits associated with the project, D represents the equivalent value of the disbenefits, and C represents the project's net cost. Similarly, the *net benefit value* (NBV) is defined as

$$\text{NBV} \equiv B - D - C \tag{8.7}$$

For a project to be desirable, $\text{BCR} > 1$ or $\text{NBV} > 0$. This rule must be applied with caution, however, since benefit quantification is usually not very precise and since the distinction between disbenefits and costs is somewhat conjectural. The BCR may vary considerably depending on whether the disbenefits are included in the numerator, or are classified as a cost and included in the denominator. If questions arise about the classification of disbenefits, it is better to use the NBV approach, because (*8.7*) gives the same value irrespective of how the disbenefits are classified.

Either present worth (the NPV), future worth, or the EUAS approach may be used to evaluate B, D, and C, provided the same method be used for all three terms.

Example 8.7 A large city is located close to a major seaport. It has been proposed that a new superhighway be built between the city and the seaport, running parallel to the present congested, two-lane highway. A group of consulting engineers has estimated that the new highway will provide the following direct benefits: (1) additional commerce between the city and the seaport, having a value of \$50 million per year; (2) future economic growth within the region over a 10-year period, resulting in an increase of \$5 million per year in commercial activity, beginning in the second year; (3) a reduction in highway accidents, resulting in a direct savings of approximately \$0.8 million per year. On the other hand, the following disadvantages or disbenefits are associated with the new highway: (i) the destruction of valuable farmland that currently contributes \$1.3 million per year to the regional economy; (ii) a decrease in commercial activity along the present highway, resulting in a loss of \$0.7 million per year. Assess the desirability of the proposed superhighway, based on a construction cost of \$280 million and a yearly maintenance cost of \$1.5 million. Assume a lifetime of 30 years and an interest rate of 7%, compounded annually.

Over the entire 30-year period, the yearly net benefits, $B - D$, are given by the EUAS method as

$$\begin{aligned} B - D &= \$50 + \$5(A/G, 7\%, 10)(P/A, 7\%, 10)(A/P, 7\%, 30) + \$0.8 - (\$1.3 + \$0.7) \\ &= \$60.0 \text{ million} \end{aligned}$$

Similarly, the yearly costs are given by

$$C = \$280(A/P, 7\%, 30) + \$1.5 = \$24.1 \text{ million}$$

and the benefit-cost ratio is

$$\text{BCR} = \frac{\$60.0}{\$24.1} = 2.49$$

Since BCR > 1, the proposed highway is considered to be desirable.

Solved Problems

8.1 Compute the net present value (NPV) of the following cash flows:

End of Year	0	1	2	3	4	5
Cash Flow, \$1000	−6	4	2	−3	−2	3

The interest rate is 15% per year, compounded annually.

By (*8.1*),

$$\begin{aligned} \text{NPV} &= -\$6000 + \$4000(P/F, 15\%, 1) + \$2000(P/F, 15\%, 2) \\ &\quad - \$3000(P/F, 15\%, 3) - \$2000(P/F, 15\%, 4) + \$3000(P/F, 15\%, 5) \\ &= -\$6000 + \$4000(1.1500)^{-1} + \$2000(1.3225)^{-1} \\ &\quad - \$3000(1.5209)^{-1} - \$2000(1.7490)^{-1} + \$3000(2.0114)^{-1} \\ &= -\$2633.99 \end{aligned}$$

8.2 A new plant to produce steel tubing requires an initial investment of \$10 million. It is expected that after three years of operation an additional investment of \$5 million will be required; and after six years of operation, another investment of \$3 million. Annual operating costs will be \$3 million and annual revenues will be \$8 million. The life of the plant is 10 years. If the interest rate is 15% per year, compounded annually, what is the NPV of this plant?

The data imply a level cash flow of

$$-\$3\,000\,000 + \$8\,000\,000 = +\$5\,000\,000$$

per year in years 1 through 10, plus flows of $-\$5\,000\,000$ and $-\$3\,000\,000$ in years 3 and 6, respectively. Hence, in units of $1 million,

$$\begin{aligned} \text{NPV} &= -10 + 5(P/A, 15\%, 10) - 5(P/F, 15\%, 3) - 3(P/F, 15\%, 6) \\ &= -10 + 5(0.19925)^{-1} - 5(1.5209)^{-1} - 3(2.3131)^{-1} \\ &= 10.5096 \end{aligned}$$

or $10 509 600.

8.3 The XYZ Company is contemplating the purchase of a new milling machine. The purchase price of the new machine is $60 000 and its annual operating cost is $2 675.40. The machine has a life of seven years, and it is expected to generate $15 000 in revenues in each year of its life, What is the net present value of the investment in this machine if the interest rate is (*a*) 8% per year, (*b*) 10% per year, (*c*) 12% per year, compounded annually? Interpret your results.

In years 1 through 7, the net annual cash flow is

$$\$15\,000 - \$2675.40 = \$12\,324.60$$

(*a*)
$$\begin{aligned} \text{NPV} &= -\$60\,000 + \$12\,324.60(P/A, 8\%, 7) \\ &= -\$60\,000 + \$12\,324.60(0.19207)^{-1} = \$4167.23 \end{aligned}$$

(*b*)
$$\begin{aligned} \text{NPV} &= -\$60\,000 + \$12\,324.60(P/A, 10\%, 7) \\ &= -\$60\,000 + \$12\,324.60(0.20541)^{-1} = \$0 \end{aligned}$$

(*c*)
$$\begin{aligned} \text{NPV} &= -\$60\,000 + \$12\,324.60(P/A, 12\%, 7) \\ &= -\$60\,000 + \$12\,234.60(0.21912)^{-1} = -\$3754.11 \end{aligned}$$

When the interest rate is less than 10%, the present worth of the annual cash flows of $12 324.60 for 7 years is greater than the $60 000 investment; hence, the NPV is a positive number. When the interest rate is 10%, the present worth of the annual cash flows is just equal to the $60 000 investment, and NPV = 0. When the interest rate is greater than 10%, the present worth of the annual cash flows is less than the investment, and the NPV is negative. Thus, when the interest rate is above 10%, it would *not* be economical to purchase the milling machine.

8.4 Determine the rate of return (ROR) for the machine of Problem 8.3.

By definition, the ROR is the interest rate at which NPV = 0; thus, by Problem 8.3(*b*), ROR = 10%. (For the given cash flows, we know that the ROR is unique.)

8.5 Find the ROR for cash flows of $-\$50\,000$ in year 0 and $+\$16\,719$ each year in years 1–5.

From (*8.2*),

$$0 = -\$50\,000 + \$16\,719(P/A, i^*\%, 5)$$

$$(A/P, i^*\%, 5) = \frac{\$16\,719}{\$50\,000} = 0.33438$$

Locating the combination $n = 5$, $A/P = 0.33438$ in Appendix A, we see that $i^* = 20\%$.

8.6 Solve Problem 8.5 by trial and error, using (*8.2*).

Try i = 15%

$$\text{NPV} = -\$50\,000 + \$16\,719(P/A, 15\%, 5) = -\$50\,000 + \$16\,719(0.29832)^{-1} = +\$6043.85$$

The given cash flows are such as to make the NPV monotonically decreasing in i; hence i^*, the value at which the NPV vanishes, must be greater than 15%.

Try i = 25%

$$\text{NPV} = -\$50\,000 + \$16\,719(P/A, 25\%, 5) = -\$50\,000 + \$16\,719(0.37185)^{-1} = -\$5038.32$$

Since the NPV now is negative, it must be that $15\% < i^* < 25\%$. Try a rate that is halfway between these two rates:

Try i = 20%

$$\text{NPV} = -\$50\,000 + \$16\,719(P/A, 20\%, 5) = -\$50\,000 + \$16\,719(0.33438)^{-1} = \$0$$

Hence, $i^* = 20\%$.

8.7 Solve Problem 8.5 by linear interpolation between $i = 15\%$ and $i = 25\%$.

Using the NPV-values calculated in Problem 8.6,

$$i^* \approx 15\% + \frac{0 - \$6043.85}{-\$5038.32 - \$6043.85}(25\% - 15\%) = 20.45\%$$

The value 20.45% is slightly in error because we have used *linear* interpolation over a relatively wide range of values (from 15% to 25%), whereas (*8.2*) is a nonlinear equation.

8.8 Compute the ROR for the following cash flows:

End of Year	0	1	2	3
Cash Flow, $1000	−50	30	−1	30

Descartes' rule tells us to expect no more than three positive values for i^*. Adding and subtracting $31 000 from CF_2, we obtain

$$0 = -\$50\,000 + \$30\,000(P/A, i^*\%, 3) - \$31\,000(P/F, i^*\%, 2)$$

As a first approximation, neglect the last term on the right:

$$(A/P, i^*\%, 3) = \frac{\$30\,000}{\$50\,000} = 0.60000$$

From the tables in Appendix A, we see that:

i	$(A/P, i\%, 3)$
30%	0.55063
40%	0.62936

Hence, for the approximate equation, $30\% < i^* < 40\%$. Restoring the neglected term should pull i^* closer to 30%; hence, try $i = 30\%$:

$$\text{NPV} = -\$50\,000 + \$30\,000(0.55063)^{-1} - \$31\,000(1.6900)^{-1} = -\$13\,860.15$$

Under the previous approximation, $\text{NPV} = -\$31\,000(P/F, i^*\%, 2) \approx -\$18\,000$, and under the present approximation, $\text{NPV} = -\$13\,860.15$; we conclude that $i^* < 30\%$. Try $i = 8\%$:

$$\text{NPV} = -\$50\,000 + \$30\,000(0.38803)^{-1} - \$31\,000(1.1664)^{-1} = +\$736.11$$

Hence, $i^* > 8\%$. Try $i = 10\%$:

$$\text{NPV} = -\$50\,000 + \$30\,000(0.40211)^{-1} - \$31\,000(1.2100)^{-1} = -\$1013.38$$

Hence, $8\% < i^* < 10\%$. Try $i = 9\%$:

$$NPV = -\$50\,000 + \$30\,000(0.39505)^{-1} - \$31\,000(1.1881)^{-1} = -\$152.33$$

Hence, $8\% < i^* < 9\%$. By linear interpolation,

$$i^* \approx 8\% + \frac{0 - \$736.11}{-\$152.33 - \$736.11}(9\% - 8\%) = 8.83\%$$

It is not difficult to show that there are no other positive values—in fact, no other real values—of i^* besides $i^* \approx 8.83\%$.

8.9 Determine the payback period (PBP) for the cash flows of (*a*) Problem 8.5, (*b*) Problem 8.8. (*c*) Comment on the results in the light of the corresponding ROR-values.

(*a*)
$$PBP = \frac{\$50\,000}{\$16\,719 \text{ per year}} = 2.99 \text{ years}$$

(*b*) We have $CF_1 + CF_2 = \$29\,000$, $CF_1 + CF_2 + CF_3 = \$59\,000$; hence,

$$PBP \approx 2 + \frac{\$50\,000 - \$29\,000}{\$59\,000 - \$29\,000}(3 - 2) = 2.7 \text{ years}$$

(*c*) The cash flows of Problem 8.5 have a *longer* payback period, but a *higher* rate of return, than those of Problem 8.8. We might say that the PBP and the ROR give opposite rankings of the two investments. However, we must always bear in mind that the PBP takes no account of interest.

8.10 In Example 8.7, assume that the affected farmers are lobbying for relocation payments and subsidies to compensate them for the lost farmland. Estimates of "equitable" payments vary widely, but one number being considered is an annual payment of $3.25 million per year to the farmers over the 30-year period. Would the project still be desirable, with this new disbenefit?

$$BCR = \frac{\$60.0 - \$3.25}{\$24.1} = 2.35$$

Since BCR > 1, the project is still desirable.

8.11 A public works project is proposed that has total present-worth benefits of $75 million and total present-worth costs of $55 million. In deliberating this proposal, some members of the town council have suggested that the project has a total present-worth disbenefit of $15 million; other members feel the $15 million should be treated as a cost. How should the proposal be evaluated?

In this particular case, it makes very little difference whether the $15 million is classified as a disbenefit or as a cost. If the $15 million is treated as a disbenefit,

$$BCR = \frac{\$75 - \$15}{\$55} = 1.09$$

If it is treated as a cost,

$$BCR = \frac{\$75}{\$55 + \$15} = 1.07$$

However, to eliminate any confusion, (*8.7*) should be used:

$$NBV = \$75 - \$15 - \$55 = \$5 \text{ million}$$

Since NBV > 0, the project is economically acceptable.

8.12 The ABC Company is considering the purchase of a new sanding machine; machine models A and B are available. Both models have a five-year life, and their cash flows are as given in Table 8-2. The interest rate is 10% per year, compounded annually. Which model should ABC buy?

Table 8-2

End of Year	Cash Flows	
	Model A	Model B
0	−$30 000	−$30 000
1	10 000	30 000
2	10 000	5 000
3	10 000	3 000
4	10 000	2 000

By the net present value method:

$$\text{NPV}_A = -\$30\,000 + \$10\,000(P/A, 10\%, 4) = -\$30\,000 + \$10\,000(0.31547)^{-1} = \$1698.74$$

$$\text{NPV}_B = -\$30\,000 + \$30\,000(P/F, 10\%, 1) + \$5000(P/F, 10\%, 2) + \$3000(P/F, 10\%, 3) + \$2000(P/F, 10\%, 4) = \$5024.93$$

By the payback period method:

$$\text{PBP}_A = \frac{\$30\,000}{\$10\,000 \text{ per year}} = 3 \text{ years} \qquad \text{PBP}_B = 1 \text{ year}$$

By the rate of return method:

for model A $$0 = -\$30\,000 + \$10\,000(P/A, i^*\%, 4)$$

$$(A/P, i^*\%, 4) = 0.33333$$

and by linear interpolation in Appendix A,

$$i^* \approx 12\% + \frac{0.33333 - 0.32923}{0.35027 - 0.33333}(15\% - 12\%) = 12.24\%$$

for model B $$0 = -\$30\,000 + \$30\,000(P/F, i^*\%, 1) + \$5000(P/F, i^*\%, 2) + \$3000(P/F, i^*\%, 3) + \$2000(P/F, i^*\%, 4)$$

By trial and error:

i^*	NPV
20%	$1172.84
25%	−$435.48

and by linear interpolation,

$$i^* \approx 20\% + \frac{\$0 - \$1172.84}{-\$435.48 - \$1172.84}(25\% - 20\%) = 23.65\%$$

It is seen that all three methods rate model B above model A; ABC should purchase model B.

Supplementary Problems

8.13 A new plant to produce tractor gears requires an initial investment of $10 million. It is expected that a supplemental investment of $4 million will be needed every 3 years to update the plant. The plant is expected to start producing gears 2 years after the initial investment is made (at the start of the third year). Revenues of $5 million per year are expected to begin to flow at the start of the fourth year. Annual operating and maintenance costs are expected to be $2 million per year. The plant has a 15-year life. List the annual cash flows.
Ans. $CF_0 = -\$10\,000\,000$, $CF_1 = CF_2 = 0$, $CF_3 = -\$6\,000\,000$, $CF_4 = CF_5 = CF_7 = CF_8 = CF_{10} = CF_{11} = CF_{13} = CF_{14} = \$3\,000\,000$, $CF_6 = CF_9 = CF_{12} = CF_{15} = -\$1\,000\,000$

8.14 What is the NPV of the plant in Problem 8.13 if the interest rate is 10% per year, compounded annually? *Ans.* −$5 336 645.33

8.15 Is the plant described in Problems 8.13 and 8.14 an economically acceptable investment?
Ans. No, because the NPV is negative.

8.16 A different plant from the one described in Problem 8.13 can be built for an initial investment of $13 million and no supplemental investments. All other data are the same as in Problems 8.13 and 8.14. (*a*) Compute the net present value. (*b*) Is this plant an economically acceptable investment?
Ans. (*a*) +$855 708.47; (*b*) yes

8.17 Is the investment described in Problem 8.16 still economically acceptable if the interest rate is 15% per year, compounded annually? Use the net present value method.
Ans. No: $NPV = -\$3\,624\,238.52 < 0$.

8.18 Compute the NPV of an investment with $CF_0 = -\$50\,000$ and $CF_j = +\$12\,000$ $(j = 1, \ldots, 6)$ if the annual interest rate, compounded annually, is (*a*) 8%, (*b*) 10%, (*c*) 12%, (*d*) 15%. (*e*) Interpret the results.
Ans. (*a*) $5473.37; (*b*) $2262.53; (*c*) −$663.98; (*d*) −$4586.74. (*e*) The investment is not economically acceptable when the interest rate is 12% or greater, in which case the present worth of the cash flows is less than the (present worth of the) investment.

8.19 Is the conclusion of Problem 8.15 changed if the interest rate is 5% per year, compounded annually? *Ans.* No: $NPV = -\$1\,928\,607.02$.

8.20 What is the NPV of the investment described in Problem 8.13 if the interest rate is 3% per year, compounded annually? *Ans.* $48 465.06

8.21 What can be said about the ROR of the plant of Problem 8.13, in view of the results of Problems 8.19 and 8.20? *Ans.* There is at least one value of i^* between 3% and 5%.

8.22 Approximate the ROR for Problem 8.18 by interpolation between the results of Problem 8.18(*b*) and (*c*). *Ans.* $i^* \approx 10.454\%$

8.23 Compute the payback period for the investment of (*a*) Problem 8.3, (*b*) Problem 8.16, (*c*) Problem 8.18. *Ans.* (*a*) PBP = 11 years; (*b*) PBP = 8 years; (*c*) PBP = 4.17 years

8.24 What is the NPV of the plant described in Problem 8.16, if the interest rate is 12% per year, compounded annually? *Ans.* −$1 195 881.54

8.25 What is the ROR of the plant described in Problem 8.16? Solve by interpolation, using the NPV data from Problems 8.16 and 8.24. *Ans.* 10.834%

8.26 What can be said about the ROR for a set of *positive* cash flows?
Ans. Since the NPV is positive for every positive *i*, no ROR exists.

8.27 Rework Problem 8.12 if the interest rate is (*a*) 15% per year, (*b*) 25% per year, compounded annually.
Ans. (*a*) $NPV_A = -\$1450.60$, $NPV_B = \$2983.17$ (buy model B). (*b*) Neither model is economically acceptable; both have negative net present values.

8.28 A new highway has been proposed to join two cities, at a total construction cost of $700 000 000. The new highway has a 20-year life. It would render obsolete the current railroad system connecting the two cities, which would be dismantled at a cost of $100 000 000. This would put the 4000 railroad employees out of work; they would each be paid $6000 per year in damages, for a total of 20 years. The railroad would require a $1 000 000 annual maintenance program if it is kept. The railroad property has an assessed valuation of $30 000 000; the property would be purchased at that figure and used as the new roadbed. The highway is estimated to yield, in taxes on the trucks using it, $0.005 per ton-mile more than the railroad; a total of 500 million annual ton-miles of use is expected. It is also estimated that the general tax revenues would increase by $10 000 each year because of the new highway. On the other hand, it is estimated that the new highway would cost $2 000 000 per year to maintain. What is the BCR for the new highway, relative to the current railroad, assuming the project would be financed by 7% per year interest-bearing bonds, with the interest compounded annually? What is the NBV? [*Hint*: The annual cost for 20 years is given by

$$(\$700\,000\,000 + \$100\,000\,000 + \$30\,000\,000)(A/P, 7\%, 20) + (\$2\,000\,000 - \$1\,000\,000) + (4000)(\$6000)$$

and the corresponding annual benefit is $(500\,000\,000)(\$0.005) + \$10\,000(A/G, 7\%, 20)$.]
Ans. BCR = 0.0249, NBV = −$100 770 537.00

8.29 Should the highway in Problem 8.28 be built? Why?
Ans. No, because the BCR is less than 1.0 and the NBV is negative, relative to keeping the railroad.

8.30 A state agency is contemplating giving a total of $5 000 000 in grants to various universities. These grants would be paid out in installments of $500 000 per year over a 10-year period. The grants would enable low-skilled persons to be retrained for new jobs, with a resulting benefit of $5000 per year in increased income for each of 1000 persons in the regional labor force, over the next 10 years. These state grants would enable the universities to obtain a total of $1 000 000 in matching federal funds for their operations, thereby reducing by $1 000 000 the total amount of state funds which would normally be required by the universities. However, the grants would require 10 persons to be added to the university staffs at an average annual salary of $20 000 each, which would have to be paid out of state funds. The current annual interest rate is 10% on funds of this type. Is the retraining program an economical investment for the state, on a present-worth basis? Compute both the present-worth BCR and the present-worth NBV. *Ans.* BCR = 8.33, NBV = +$27 035 330.26; yes, it is economical.

8.31 Does the answer to Problem 8.30 change if the BCR and NBV are computed in terms of annualized costs? *Ans.* BCR = 8.33, NBV = +$4 400 000; again the investment is economical.

8.32 A state agency is contemplating building a new 5630-acre industrial park. The property can be acquired at a cost of $1000 per acre. Roadways and other improvements are estimated to cost a total of $1000 per acre, with these costs spread evenly over the next 10 years. The few residents currently on the property will be moved out over the next three years; the displacement costs are estimated at $1 000 000 this year, $500 000 next year, and $200 000 the year after. The park is expected to provide the state with new tax revenues of $5 000 000 per year, starting five years from now, with an increase of $2 000 000 per year thereafter. The state has decided to evaluate this project on the basis of a 10-year lifetime. The funds for the project will be borrowed at an interest rate of 7% per year, compounded annually. (*a*) What is the present-worth NBV of the project? (*b*) What is the present-worth BCR? (*c*) Is the project economically justifiable? *Ans.* (*a*) $14 401 566; (*b*) 2.30; (*c*) yes

Chapter 9

Choosing Among Investment Alternatives

9.1 SETTING THE MARR

In Chapter 6 we introduced the MARR as the smallest yield rate at which a proposed investment would be acceptable. How is this cut-off rate arrived at?

1. The MARR may be set equal to the interest rate that is available at a local savings bank or other institution. The MARR then becomes the "opportunity cost of money," in that it measures the opportunity lost from not placing money in the bank.
2. For most businesses, the savings bank rate would be lower than their usual overall rate of return on investment. Thus, the MARR is sometimes set equal to the firm's current average return on total investment.
3. The MARR may be purposely set higher than either the bank rate on savings or the firm's current return on investment. It may be set according to the firm's long-range profit goals, so as to achieve a desired future growth rate; it may be set at a high level to encourage the search for more profitable new ventures; it may be chosen large to offset the high degree of risk attached to the investment.

Under option 1 and (within the law of averages) under option 2, the MARR is an *attainable* rate; i.e., there are investment alternatives that actually achieve that rate. However, under option 3, the MARR is a *target* rate, with no guaranteed means of realization.

Example 9.1 The ABC Company is currently earning an average before-tax return of 25% on its total investment. The board of directors of ABC is considering three proposals as given in Table 9-1.

Table 9-1

	Cash Flows		
End of Year	Proposal A	Proposal B	Proposal C
0	−$40 000	−$60 000	−$50 000
1	18 000	25 000	27 000
2	18 000	25 000	27 000
3	18 000	25 000	27 000
4	18 000	25 000	27 000

Proposal A is for a new machine that will replace one of their older, worn-out pieces of equipment; this machine is vital to ABC's production. Proposal B is for a plant expansion. Proposal C is for an addition to ABC's product line. There is a high probability that this product could fail in the marketplace, resulting in the loss of most of the $50 000 initial investment. The board feel that they would need at least a 40% rate of return on this project to compensate for its additional riskiness. Which of these three proposals are acceptable?

Compute net present values, using the attainable MARR of 25% for proposals A and B, and the target MARR of 40% for proposal C:

$$\text{NPV}_A = -\$40\,000 + \$18\,000(P/A, 25\%, 4) = +\$2508.97$$
$$\text{NPV}_B = -\$60\,000 + \$25\,000(P/A, 25\%, 4) = -\$959.76$$
$$\text{NPV}_C = -\$50\,000 + \$27\,000(P/A, 40\%, 4) = -\$71.19$$

Only proposal A is acceptable ($\text{NPV}_A > 0$).

The ROR method leads to the same conclusion: linear interpolation in Appendix A gives

$$i_A^* \approx 28.75\% \qquad i_B^* \approx 24.07\% \qquad i_C^* \approx 39.99\%$$

and only i_A^* exceeds the associated MARR.

Because NPV_B and NPV_C, though negative, are small in magnitude (causing i_B^* and i_C^* to be just under their associated MARRs), the ultimate decisions concerning proposals B and C may have to be made on the basis of other considerations, such as ABC's long-term product strategies and the company's ability to raise capital. If capital is scarce, proposal A must be given the highest priority, since ABC's continued profits appear to depend on that piece of machinery.

Example 9.2 The XYZ Company has $50 million which can be invested in proposal A ($i_A^* = 17\%$) or in proposal B ($i_B^* = 29\%$); or else it can exercise the *do-nothing alternative* and invest the $50 million in modernizing current operations. A target MARR of 35% has been established by XYZ's management to achieve their long-range plans and strategies. The XYZ Company currently earns an average of 25% on its total investment in plant and equipment, some of which is very old. Which alternative should XYZ pursue?

For the do-nothing alternative, $i^* = 25\%$; thus, none of the three alternatives meets the desired 35% MARR. If the company is serious about the 35% MARR, then additional alternatives should be sought. Proposal B, which has an ROR slightly better than the current average rate of return on total investment, would clearly enhance the company's average rate of return. In addition, given that some of the company's plant and equipment is "very old," using the available $50 million for refurbishing this old plant and equipment might also improve the company's average rate of return. One reasonable strategy would be to spend part of the $50 million on new plant and equipment, and then to search for higher-profit proposals (e.g., a 35% ROR) on which to spend the balance.

9.2 PROJECT SELECTION AND BUDGET ALLOCATION

Determining the best way to allocate a given budget among several competing projects is a commonly encountered problem, because often there are more worthwhile project proposals and ideas than can be funded with the available monies. The solution principle is to evaluate each project in terms of present worth or some similar measure, and to choose that set of projects for which the sum of the measures is a maximum, subject to the budget constraint.

Independent Projects

Two or more proposals or projects are *independent* when the acceptance or rejection of any one of them does not entail the acceptance or rejection of any other. For instance, a proposal to air-condition the company offices and a proposal to undertake an advertising campaign for a new product would usually be considered independent.

For independent projects, the following selection algorithm will always maximize the financial return on the available monies.

Step 1 Compute i^* for each project.

Step 2 Eliminate any project whose i^*-value is less than the MARR (if no MARR exists, omit this step).

Step 3 Arrange the surviving proposals from step 2 in descending order of i^*-value.

Step 4 Select proposals from the top of this list downward, until an additional selection would exceed the available funds or the budget.

If, as often will be the case, some funds remain at the end of step 4, there are three options: (i) if one or more of the remaining projects is divisible into subprojects, then these subprojects may be funded, using the above algorithm, until the available funds are exhausted; (ii) the remaining funds may be invested in the do-nothing alternative, at the MARR (for an attainable MARR) or at some rate less than the MARR (for a target MARR); (iii) the remaining funds are simply "left over."

Example 9.3 The BK Company is considering five proposals for new equipment, as indicated in Table 9-2. Each piece of equipment has a life of 100 years. Treating that period as infinite, the ROR will be the interest rate at which I is the capitalized equivalent of the perpetual series of payments R; hence, (*7.8*) gives the third row of Table 9-2. The BK Company has established a MARR of 11% and has a budget of $325 000. Which proposal(s) should the company select?

Table 9-2

	Proposal 1	Proposal 2	Proposal 3	Proposal 4	Proposal 5
Annual Revenue, R	$5 000	$6 000	$25 000	$16 000	$20 000
Investment, I	$60 000	$50 000	$100 000	$100 000	$100 000
$i^* \approx R/I$	$8\frac{1}{3}$%	12%	25%	16%	20%

Using the selection algorithm, we obtain the following list:

	i^*	Investment	
Proposal 3	25%	$100 000	
Proposal 5	20%	$100 000	
Proposal 4	16%	$100 000	
- - - - - - - -	- - - -	- - - - - -	Budget cut-off
Proposal 2	12%	$50 000	
- - - - - - - -	- - - -	- - - - - -	MARR cut-off
Proposal 1	$8\frac{1}{3}$%	$60 000	

Proposal 2 is acceptable from the standpoint of the MARR criterion, but insufficient funds are available to include it. Thus, proposals 3, 5, and 4 are selected, and $25 000 is left unspent from the $325 000 budget.

Mutually Exclusive Projects

A set of projects are *mutually exclusive* if at most one of them may be accepted. It is thus a question of picking the single economically best project (or of rejecting them all).

For mutually exclusive projects, the selection algorithm given below will always yield the maximum total return on the total amount invested. First, we shall need some terminology. Let I denote the investment cost of a project, and R the measure of revenues (present worth, EUAS, etc.) from the project. We shall say that project 1 *dominates* project 2 if $I_1 \leq I_2$ and $R_1 \geq R_2$. Clearly, a dominated project can never be the best of a mutually exclusive set. Further, for any two projects—a *standard* and a *challenger*—define the *incremental rate of return* of the challenger as

$$\Delta i^* \equiv \frac{R_{\text{challenger}} - R_{\text{standard}}}{I_{\text{challenger}} - I_{\text{standard}}}$$

Step 1 Eliminate any project whose investment exceeds the budget.

Step 2 Arrange the surviving projects in ascending order of investment (break any investment-ties arbitrarily). Now eliminate any project that is dominated by another project; the candidates that remain will be in ascending order both of investment and of return. Compute i^* for each candidate.

Step 3 Eliminate from further consideration any candidate having $i^* <$ MARR.

Step 4 From the surviving candidates, select as the standard that candidate which has the smallest investment.

Step 5 Compute the incremental rate of return of the challenger that immediately succeeds the standard in the list of candidates.

Step 6 If $\Delta i^* \leq$ MARR, eliminate this challenger from further consideration and repeat step 5 for the next challenger; if $\Delta i^* >$ MARR, eliminate the old standard from further consideration, replace it with this challenger as the new standard, and repeat step 5.

Step 7 Select the one surviving candidate: it is the best alternative.

Example 9.4 The KLN Company is attempting to determine the economically best size of processor machine for their facilities. The six alternative machine sizes which are feasible are as given in Table 9-3. Each machine has a life of 100 years and no salvage value, so that $i^* \approx R/I$, as in Example 9.3. The company has a total capital budget of \$350 000 and a MARR of 15%. Which machine should they buy?

Table 9-3

Size of Machine	Annual Revenue, R	Investment, I	i^*
Economy	\$ 7 200	\$ 60 000	12%
Regular	25 000	100 000	25%
Super	36 000	200 000	18%
Delux	45 000	220 000	20.45%
Bulk	50 000	300 000	16.67%
Extended	52 000	385 000	20.5%

The Extended machine is unacceptable according to step 1 of the selection algorithm, and the Economy machine is unacceptable according to step 3. The application of steps 5 and 6 to the four surviving candidates is shown in Table 9-4; step 7 gives Delux as the winner.

Table 9-4

Comparisons	i^*	Steps 5 and 6
Regular vs. Super	25% vs. 18%	Standard #1 vs. Challenger #1: $\Delta i^* = \dfrac{\Delta R}{\Delta I} = \dfrac{\$36\,000 - \$25\,000}{\$200\,000 - \$100\,000} = \dfrac{\$11\,000}{\$100\,000} = 11\% < \text{MARR}$ **Decision:** reject challenger #1 and repeat step 5
Regular vs. Delux	25% vs. 20.45%	Standard #1 vs. Challenger #2: $\Delta i^* = \dfrac{\$20\,000}{\$120\,000} = 16.7\% > \text{MARR}$ **Decision:** replace standard #1 (Regular) with challenger #2 (Delux) and repeat step 5
Delux vs. Bulk	20.45% vs. 16.67%	Standard #2 vs. Challenger #3: $\Delta i^* = \dfrac{\$5\,000}{\$80\,000} = 6.25\% < \text{MARR}$ **Decision:** reject challenger #3

Let us examine the logic of the selection algorithm, on the assumption that the company can realize 15% (the MARR) by implementing the do-nothing alternative. Consider the first comparison in Table 9-4. Super costs ΔI = \$100 000 more than Regular, and yields ΔR = \$11 000 more per year. If the company chose Super, it would, in effect, be making \$11 000 a year on a \$100 000 investment; that is, it would be investing at rate $\Delta i^* = 11\%$, whereas it could be earning MARR = 15%. Choosing Super would thus entail an opportunity loss of

$$(15\% - 11\%)(\$100\,000) = \$4000 \text{ per year}$$

Or, looked at in a slightly different way, if the Regular machine is purchased, at a saving of \$100 000, the company will earn \$25 000 a year on the machine, plus 15% × \$100 000 = \$15 000 a year on the do-nothing alternative. This is a total annual return of \$40 000 on a total investment of \$200 000. The same total investment in the Super machine will earn only \$36 000 a year.

In this first comparison, it so happens that the economically superior machine has the larger i^*-value. Note however, that the eventual winner, Delux, has a *smaller* i^*-value than Regular. As we have seen, when purchase prices differ, a mere comparison of i^*-values is not decisive; one must also consider what will be done with any funds left over from the purchase of the cheaper machine.

Other Interrelationships Between Projects

A project may be *contingent* upon some other project, in the sense that the acceptance or rejection of one may result in the corresponding acceptance or rejection of the other. For example, the purchase of a new computer storage disk may be contingent on the purchase of a new computer.

Some projects may be *joint*, in that either both are accepted or both are rejected. For instance, though they may be purchased separately, the tractor and trailer of a rig for highway hauling of steel are normally joint items.

Some projects may be *financially interdependent*; i.e., approval of one exhausts the available funds and thus precludes approval of the other.

Joint projects should be treated as a single, total project or investment. Contingent projects should be evaluated both jointly and separately. The basic investment should first be evaluated alone. The contingent items should then be brought in and their effect on the total investment evaluated. Financial interdependences are usually resolved by considering the *irreducibles*, those factors to which a dollar value cannot be attached.

Example 9.5 Which project(s) in Table 9-5 should be approved, if the budget is \$150 000 and the MARR is 15%?

Table 9-5

Project	Investment	i^*
A	\$100 000	20%
B	50 000	20%
C	50 000	20%
D	50 000	20%
E	150 000	20%

This is an instance of financial interdependence. Because of the \$150 000 budget, selecting project A and either B, C, or D precludes selecting any others; selecting E precludes selecting any others; and selecting B and C and D precludes selecting any others. Thus, there are five alternatives (investment portfolios), each representing a total investment of \$150 000 and each with an ROR of 20%. The choice among them must be made on the basis of the intrinsic characteristics of the projects, the need for a diversified portfolio, and other irreducible factors.

9.3 THE REINVESTMENT FALLACY

It is implicitly assumed in the NPV, EUAS/EUAC, and ROR methods that any cash inflows generated by an investment are reinvested, at the rates MARR, MARR, and i^*, respectively. If such an assumption does not hold, and if the assets being compared have unequal service lives, fallacious results may be obtained.

Example 9.6 Consider two competing projects, for which MARR = 16%:

	C	A	n
Project A	\$100 000	\$23 000	9
Project B	100 000	35 000	4

Here, C is the initial investment, A is the annual net cash inflow, and n is the service life of the asset. The ROR method yields:

project A

$$0 = -\$100\,000 + \$23\,000\,(P/A, i^*, 9)$$

$$(A/P, i^*, 9) = 0.23$$

$$i^* \approx 17.7\% \text{ (by interpolation in Appendix A)}$$

project B

$$0 = -\$100\,000 + \$35\,000(P/A, i^*, 4)$$

$$(A/P, i^*, 4) = 0.35$$

$$i^* \approx 15\%$$

Hence, according to the MARR, project A is acceptable and project B is not.

However, suppose that the cash flows can be reinvested at 25%, compounded annually. Thus, the \$23 000 annual cash inflows from project A are actually equivalent to a future value

$$\$23\,000(F/A, 25\%, 9) = \$593\,400$$

nine years hence, and the annual cash inflows from project B are actually equivalent to a future value

$$\$35\,000(F/A, 25\%, 4)\,(F/P, 25\%, 5) = \$615\,900$$

nine years hence. Thus (the initial investments being equal) project B is actually the preferred alternative.

The reinvestment fallacy can be avoided if the MARR is set at the reinvestment rate (whose value, however, may be very difficult to predict) and if future values based on this MARR are compared, as in Example 9.6. As for the matter of unequal lives, it can sometimes be ignored, and the NPV, EUAC, or ROR method applied notwithstanding. For other situations, a replacement method which assumes that each asset, at the end of its useful life, is replaced with a new asset identical in kind, may be more appropriate. This method will be employed in Chapter 10.

Solved Problems

9.1 The management of the Conway Corporation is considering five alternative new-product proposals that their employees have submitted to them:

Proposal	Rate of Return
Fryer	49%
Box Loader	26%
Conveyor	19.5%
Planer	23%
Cutter-Loader	26.5%

Conway currently enjoys an average return of 26% on total investment. Which proposal(s) is (are) acceptable?

Using the current average return on total investment as the MARR, only Fryer and Cutter-Loader are *strictly* acceptable. If the rule is: ROR $\geq$ MARR, then Box Loader is also acceptable.

9.2 Rework Problem 9.1 if Conway desire a future average return of 31%.

Only Fryer is acceptable under the target MARR.

9.3 The Wyandot Company currently earns an average rate of return of 30% on its total investment. The board of directors of Wyandot is considering the three proposals whose cash flows are specified in Table 9-6. Which proposal(s) is (are) acceptable, if the board of directors has set a target MARR of 25%? Make an NPV calculation.

Table 9-6

End of Year	Proposal A	Proposal B	Proposal C
0	−$50 000	−$75 000	−$100 000
1	15 000	30 000	35 000
2	15 000	30 000	35 000
3	15 000	30 000	35 000
4	15 000	30 000	35 000

$$NPV_A = -\$50\,000 + \$15\,000\,(P/A, 25\%, 4) = -\$14\,575.85$$
$$NPV_B = -\$75\,000 + \$30\,000\,(P/A, 25\%, 4) = -\$4151.71$$
$$NPV_C = -\$100\,000 + \$35\,000\,(P/A, 25\%, 4) = -\$17\,344.00$$

Since all three net present values are negative, none of the projects is acceptable.

9.4 Solve Problem 9.3 by an ROR calculation.

proposal A
$$(A/P, i_A^*, 4) = \frac{\$15\,000}{\$50\,000} = 0.30000$$
$$7\% < i_A^* < 8\% \quad \text{(from Appendix A)}$$

proposal B
$$(A/P, i_B^*, 4) = \frac{\$30\,000}{\$75\,000} = 0.40000$$
$$20\% < i_B^* < 25\% \quad \text{(from Appendix A)}$$

proposal C
$$(A/P, i_C^*, 4) = \frac{\$35\,000}{\$100\,000} = 0.35000$$
$$12\% < i_C^* < 15\% \quad \text{(from Appendix A)}$$

All three rates of return are smaller than the target MARR (25%), and so none of the projects is acceptable.

9.5 With reference to Problems 9.3 and 9.4, how would Wyandot's average return on investment be affected by the acceptance of proposal B?

The effect would depend on the amount of Wyandot's total invested capital. Suppose Wyandot to be a very small firm, whose total investment before accepting proposal B is $225 000. Since i_B^*, the rate of return of proposal B, is 21.85% (by linear interpolation in Appendix A), the average rate of return after acceptance of proposal B would be

$$\frac{(30\%)(\$225\,000) + (21.85\%)(\$75\,000)}{\$300\,000} = 27.96\%$$

or a decrease of about 2%. However, if Wyandot were a larger company, with, say, \$925 000 in invested capital, then the new average rate would be

$$\frac{(30\%)(\$925\,000) + (21.85\%)(\$75\,000)}{\$1\,000\,000} = 29.39\%$$

a decrease of only $\frac{6}{10}\%$.

9.6 Refer to Problem 9.3. Suppose that if proposal B were rejected, the competition would be likely to introduce a new product that would cut Wyandot's market share in half, and hence cause them to suffer a 50% reduction in current profits. Should proposal B be accepted or rejected under these circumstances?

We have seen that acceptance of proposal B is unprofitable *per se*. However, rejection could only prove more unprofitable (unless Wyandot's profits are minuscule to begin with). Thus, proposal B should be accepted, as "the lesser of two evils."

9.7 The Clearwater Company has a budget of \$500 000 which can be spent on the five independent projects of Table 9-7. If MARR = 20%, how should the budget be allocated?

Table 9-7

Project Number	i^*	Total Project Cost
1	29.1%	\$150 000
2	10.5%	50 000
3	21.5%	200 000
4	19.5%	75 000
5	23.2%	25 000

The selection algorithm for independent projects gives:

Step 2 Eliminate projects 2 and 4.

Step 3 Select

Project 1	\$150 000
Project 5	25 000
Project 3	200 000
TOTAL	\$375 000

with \$500 000 − \$375 000 = \$125 000 unspent.

9.8 Rework Problem 9.7 if (*a*) MARR = 25%, (*b*) MARR = 19%, (*c*) MARR = 18% and capital is rationed at \$400 000.

(*a*) Only project 1 can be funded; \$350 000 remains unspent.

(*b*) Only project 2 is unacceptable, and \$450 000 is spent as follows:

Project 1	\$150 000
Project 3	200 000
Project 4	75 000
Project 5	25 000
TOTAL	\$450 000

(*c*) The MARR eliminates project 2, and the ranking becomes:

	Project Cost	Cumulative Amount Spent
Project 1	$150 000	$150 000
Project 5	25 000	175 000
Project 3	200 000	375 000
Project 4	75 000	

Unless it is divisible, project 4 cannot be funded: its inclusion would exceed the $400 000 budget constraint.

9.9 Grampian Manufacturing Company is attempting to determine the "best"-sized milling machine for their production shop. Five alternative sizes are available, as given in Table 9-8. Grampian has a budget of $250 000, and MARR = 15%. Which size machine should they purchase? Assume that $n = 100$ years and that the ultimate salvage value is zero for each machine.

Table 9-8

Size	Annual Revenue	Initial Cost	i^*
Economy	$ 5 000	$ 50 000	10%
Regular	25 000	100 000	25%
Super	36 000	200 000	18%
Delux	45 000	220 000	20.45%
Super Delux	50 000	300 000	16.67%

This situation involves mutually exclusive projects. The selection algorithm (Section 9.2) gives:

Step 1 Eliminate Super Delux.

Step 2 See Table 9-8.

Step 3 Eliminate Economy.

Steps 4 through 6 Compare Super against Regular:

$$\Delta i^* = \frac{36\,000 - 25\,000}{200\,000 - 100\,000} = 11\% < \text{MARR}$$

hence, eliminate Super. Compare Delux against Regular:

$$\Delta i^* = \frac{45\,000 - 25\,000}{220\,000 - 100\,000} = 16.67\% > \text{MARR}$$

hence, Delux becomes the new standard.

Step 7 Select Delux.

In this case, $30 000 will be left over from the original $250 000 budget.

9.10 Rework Problem 9.9 for MARR = 20%.

Now only Regular and Delux survive step 1 of the algorithm. From step 5, with Delux as the challenger,

$$\Delta i^* = 16.67 < \text{MARR}$$

Hence, Regular is selected, and $100 000 is spent.

9.11 Rework Problem 9.9 for a budget of \$200 000.

In this case, Delux and Super Delux are eliminated in step 1 of the algorithm, and Economy in step 3. Then, only Regular and Super remain; from Problem 9.9, Regular wins.

9.12 The CCC Corporation is weighing the purchase of a computer. The basic machine costs \$200 000; the costs of various peripheral equipment are:

Slow Printer	\$ 10 000
Fast Printer	20 000
Low-Resolution Video Display	2 000
High-Resolution Video Display	5 000
Disk Drives (each)	2 000
Remote-User Ports (each)	2 000
Software	100 000
Special Disks	1 000

The job that the computer would perform is now being done by hand, by five persons, at an annual salary and overhead cost of \$100 000. Though these people would all be replaced by the computer, the purchase of the computer would necessitate the hiring of two programmer-operators and one mathematician, at a total salary and overhead cost of \$80 000. Classify the investment decisions and proposals involved in this situation.

The peripheral equipment are all contingent projects—contingent on the purchase of the basic computer. The slow and fast printers are mutually exclusive, as are the high- and low-resolution videos. Some type of printer and/or video, some software, some disks, and one or more disk drives would seem to be mandatory: they are joint proposals with the computer. The do-nothing alternative (continuing with the handicraft technology of five persons) and the computer purchase alternative are mutually exclusive projects.

9.13 If, in Problem 9.12, CCC has only \$210 000 available for the computer system, exclusive of software, does this introduce any financial interdependences?

Yes. If \$10 000 is spent on the slow printer, then no other peripherals can be purchased. If \$10 000 is spent on some combination of peripherals (e.g., special disks plus one remote-user port plus one disk drive plus one high-resolution video), then no other items can be purchased beyond the basic computer.

9.14 For the data of Problem 9.12 and Table 9-9, and for MARR = 15%, should the computer be purchased? What peripherals should be purchased?

Table 9-9

Comparison	Δi^*
Do-Nothing vs. Basic Computer plus Software	16%
Slow vs. Fast Printer	8%
Low- vs. High-Resolution Video Computer	13%
Computer with One Disk Drive vs. Computer with Two Disk Drives	12%
Zero vs. Two Remote Ports	21%

The basic computer plus software, one slow printer, one disk drive, one low-resolution video, and two remote ports should be purchased, since $\Delta i^* > \text{MARR}$ for these items. The high-resolution video may be justifiable on the basis of irreducibles, such as operator eye-fatigue, since its Δi^* is not that far away from the established MARR.

9.15 ABC Company have decided to automate certain of their procedures by installing a computer system. The cash flows for two competing systems are given in Table 9-10; both systems have a five-year life and zero salvage value. If the MARR is 15%, which system should ABC purchase?

Table 9-10

End of Year	System 1	System 2
0	−$50 000	−$75 000
1	22 000	24 000
2	22 000	24 000
3	22 000	24 000
4	22 000	24 000
5	22 000	24 000

In the case of equal lifetimes, the NPV (or the EUAS) is a linear function of initial cost and annual revenue; hence we compute

$$\begin{aligned}\Delta\text{NPV} &\equiv \text{NPV}_2 - \text{NPV}_1\\ &= [-\$75\,000 - (-\$50\,000)] + (\$24\,000 - \$22\,000)\,(P/A, 15\%, 5)\\ &= -\$18\,295.79\end{aligned}$$

As $\Delta\text{NPV} < 0$, system 2 is economically inferior to system 1; system 1 should be purchased.

9.16 In a situation like that of Problem 9.12, the directors of a firm are trying to decide whether to buy computer system A, to buy computer system B, or to stay with the current manual technology. Advise them, given a MARR of 15%, a planning horizon of 4 years, and costs as in Table 9-11.

Table 9-11

	Manual	System A	System B
Equipment			
Computer		$200 000	$200 000
Printer		20 000	10 000
Video		5 000	2 000
Disk Drives		4 000	4 000
Remote Ports		2 000	0
Disks		5 000	2 000
		TOTAL $236 000	TOTAL $218 000
Software		100 000	50 000
Annual Manpower	$100 000	80 000	40 000
Annual Overhead	50 000	20 000	40 000

We compare either computer system to the manual technology by the difference method of Problem 9.15, choosing present-worth cost as economic parameter. For system B versus manual,

$$\Delta PW = \$268\,000 + (\$80\,000 - \$150\,000)(P/A, 15\%, 4)$$
$$= \$68\,154.17 > 0$$

and for system A versus manual,

$$\Delta PW = \$336\,000 + (\$100\,000 - \$150\,000)(P/A, 15\%, 4)$$
$$= \$193\,252.98 > 0$$

The strict conclusion is that the current manual technology should be retained. However, a consideration of the irreducibles (e.g., improved output quality when the job is done by computer) might make system B more attractive.

9.17 Illustrate the reinvestment fallacy by supposing that, in Problem 9.15, the revenues from system 2 could be reinvested at 40% in years 3 through 5.

At the end of five years, the net future worth of system 1 is:

$$FW_1 = -\$50\,000(F/P, 15\%, 5) + \$22\,000(F/A, 15\%, 5)$$
$$= -\$50\,000(2.0114) + \$22\,000(6.7424) = \$47\,762.80$$

and the net future worth of system 2 is (draw a time diagram):

$$FW_2 = -\$75\,000(F/P, 15\%, 5) + \$24\,000(F/P, 15\%, 1)(F/P, 40\%, 3) + \$24\,000(F/A, 40\%, 4)$$
$$= -\$75\,000(2.0114) + \$24\,000(1.15)(2.7440) + \$24\,000(7.1040)$$
$$= \$95\,375.40$$

The effect of the reinvestment is to reverse the conclusion of Problem 9.15: now, system 2 is the better.

Supplementary Problems

9.18 The executives of the XYZ Company are considering the three independent proposals whose cash flows are given in Table 9-12. The MARR is 10%. Evaluate these three proposals by the NPV method.

Table 9-12

End of Year	Proposal A	Proposal B	Proposal C
0	−$50 000	−$80 000	−$150 000
1	16 461.50	28 021.60	45 000
2	16 461.50	28 021.60	45 000
3	16 461.50	28.021.60	45 000
4	16 461.50	28 021.60	45 000

Ans. $NPV_A = \$2180.87$, $NPV_B = \$8824.93$, $NPV_C = -\$7355.69$; proposal C is unacceptable.

9.19 Rework Problem 9.18 for MARR = 15%.
Ans. $NPV_A = -\$3003.40$, $NPV_B = 0$, $NPV_C = -\$21\,527.68$; proposals A and C are unacceptable and proposal B is barely acceptable.

9.20 Evaluate the three proposals in Problem 9.18 using the ROR method and linear interpolation.
Ans. $i_A^* = 12\%$, $i_B^* = 15\%$, $i_C^* = 7.71\%$; since $i_C^* < \text{MARR}$, proposal C is unacceptable.

9.21 The capital budgeting committee of the ABC Company is contemplating five independent proposals for projects to be included in the forthcoming year's budget; their cash flows are given in Table 9-13. The ABC Company has established a MARR of 20%. Assuming that capital is not rationed, which projects should the company select and what is the total investment required? Use the ROR method.

Table 9-13

End of Year	Project 1	Project 2	Project 3	Project 4	Project 5
0	−$100 000	−$200 000	−$150 000	−$80 000	−$300 000
1	35 027	77 258	63 516	32 000	98 769
2	35 027	77 258	63 516	32 000	98 769
3	35 027	77 258	63 516	32 000	98 769
4	35 027	77 258	63 516	32 000	98 769

Ans. $i_1^* = 15\%$, $i_2^* = 20\%$, $i_3^* = 25\%$, $i_4^* = 21.85\%$ (by interpolation), $i_5^* = 12\%$. Projects 2, 3, and 4 should be selected, at a total investment of $430 000.

9.22 Would the results of Problem 9.21 change if (*a*) MARR = 10%? (*b*) MARR = 13% and capital is rationed at $430 000? *Ans.* (*a*) no; (*b*) no

9.23 How would the results of Problem 9.21 change if the MARR is 13%, the budget constraint is $480 000, and all the projects are divisible into smaller projects?
Ans. Select projects 2, 3, and 4, and spend $50 000 on project 1.

9.24 How would the results of Problem 9.23 change if the MARR is 16% and none of the projects are divisible? *Ans.* Select projects 2, 3, and 4, and leave $50 000 unspent.

9.25 Would the results of Problem 9.24 change if all the projects were divisible? *Ans.* no

9.26 The executives of the XYZ Company are attempting to determine the economically best process-control computer to purchase for one of their production lines. The choice has been narrowed to the five mutually exclusive alternatives whose cash flows are presented in Table 9-14. If capital is not rationed and the MARR is 7%, which computer should the company purchase? *Ans.* C or D

Table 9-14

End of Year	Computer A	Computer B	Computer C	Computer D	Computer E
0	−$20 000	−$30 000	−$28 000	−$35 000	−$25 000
1	6 309.40	9 057.60	9 218.44	11 523.05	7 886.75
2	6 309.40	9 057.60	9 218.44	11 523.05	7 886.75
3	6 309.40	9 057.60	9 218.44	11 523.05	7 886.75
4	6 309.40	9 057.60	9 218.44	11 523.05	7 886.75

9.27 Will the result of Problem 9.26 change if the MARR is (*a*) 10%? (*b*) 11%? (*c*) 13%?
Ans. (*a*) no; (*b*) no; (*c*) yes (none of the alternatives is acceptable)

9.28 Rework Problem 9.26 if capital is rationed at (*a*) $30 000, (*b*) $28 000, (*c*) $25 000.
Ans. (*a*) C; (*b*) C; (*c*) E

9.29 For what combinations of capital rationing and MARR values in Problem 9.26 would computer B be preferred to computer C? *Ans.* None: C dominates B (Section 9.2).

9.30 The executives of the ABC Company are trying to select the most economical feeder machine. Cash flows for the six available models are shown in Table 9-15. The MARR is 8%. (*a*) Compute the ROR for each alternative model. (*b*) Determine which model should be purchased.
Ans. (*a*) $i_A^* = 10\%$, $i_B^* = 25\%$, $i_C^* = 18\%$, $i_D^* = 20.5\%$, $i_E^* = 19.5\%$, $i_F^* = 20.5\%$; (*b*) F

Table 9-15

End of Year	A	B	C	D	E	F
0	−\$50 000	−\$100 000	−\$200 000	−\$220 000	−\$250 000	−\$380 000
1	13 190	37 185	63 991.20	74 386.40	82 693.50	128 485.60
2	13 190	37 185	63 991.20	74 386.40	82 693.50	128 485.60
3	13 190	37 185	63 991.20	74 386.40	82 693.50	128 485.60
4	13 190	37 185	63 991.20	74 386.40	82 693.50	128 485.60
5	13 190	37 185	63 991.20	74 386.40	82 693.50	128 485.60

9.31 Which model should be purchased in Problem 9.30, if there is a budget constraint of (*a*) \$280 000? (*b*) \$230 000? (*c*) \$210 000? *Ans.* (*a*) E; (*b*) D; (*c*) C

9.32 How would the result of Problem 9.30(*b*) change if the MARR is (*a*) 9%? (*b*) 10%? (*c*) 12%? (*d*) 15%? (*e*) 20%? *Ans.* (*a*)–(*e*) no change

9.33 How would the result of Problem 9.30(*b*) change if MARR = 15% and the budget constraint is (*a*) \$210 000? (*b*) \$280 000? *Ans.* (*a*) B (the only choice); (*b*) D

9.34 The KJL Company is contemplating five independent projects, with cash flows as in Table 9-16. The MARR is 12% and the budget constraint is \$200 000. (*a*) Compute the rate of return for each project. (*b*) What is the optimum portfolio, if the minimum desired payback period (Section 8.3) is two years?
Ans. (*a*) $i_A^* = i_B^* = i_C^* = i_D^* = i_E^* = 15\%$. (*b*) $PBP_A = PBP_B = PBP_C = 2.28$ years, $PBP_D = 2.055$ years $PBP_E = 2.0$ years; the only acceptable choice is project E.

Table 9-16

End of Year	Project A	Project B	Project C	Project D	Project E
0	−\$100 000	−\$50 000	−\$75 000	−\$60 000	−\$95 000
1	43 798	21 899	38 848.50	30 000	50 000
2	43 798	21 899	38 848.50	29 000	45 000
3	43 798	21 899	38 848.50	18 228	26 609

9.35 How would the result of Problem 9.34(*b*) change if (*a*) the minimum desired payback period is 2.2 years? (*b*) the minimum desired payback period is 2.2 years and the budget constraint is \$150 000?
Ans. (*a*) D, or E, or D and E; (*b*) D or E

9.36 Two alternative cleaning machines are being considered as replacements for an older, worn-out cleaner. The cash flows for the two mutually exclusive alternatives are presented in Table 9-17; the MARR is 10%. (*a*) Compute the present-worth difference of value (ΔPW) between the two machines. (*b*) Compute the incremental rate of return (Δi^*) on the investment difference between the two machines. (*c*) Which machine should be purchased? *Ans.* (*a*) $816.33; (*b*) 37.9%; (*c*) machine #2

Table 9-17

End of Year	Machine #1	Machine #2
0	−$20 000	−$28 000
1	4 864.60	8 419.88
2	4 864.60	8 419.88
3	4 864.60	8 419.88
4	4 864.60	8 419.88
5	4 864.60	8 419.88
6	4 864.60	8 419.88

Chapter 10

Equipment Replacement and Retirement

10.1 RETIREMENT AND REPLACEMENT DECISIONS

The decision to replace equipment or to retire it (to take the equipment out of service without replacing it) can be motivated by the physical impairment of the equipment, its obsolescence, or external economic conditions. Retirement and replacement decisions should always be based on economics rather than on whether or not the equipment has reached the end of its physical service life. A piece of equipment may have many years of service life remaining beyond the point at which it has become uneconomical to operate it.

All past investments and expenses connected with the equipment are *sunk costs*, which do not enter into a retirement/replacement decision. *Only current and future costs and investments are relevant.*

10.2 ECONOMIC LIFE OF AN ASSET

As operating equipment ages, the usual pattern is for its capital costs to decline while its operating costs rise. When summed, these two cost functions often result in a cost function that is generally U-shaped. Ideally, the equipment should be retired at the lowest point on this total cost function.

Example 10.1 A machine has an initial cost of \$10 000. Being a special-purpose custom-built unit, it can only be resold as scrap at \$500, no matter what its age. The machine has a 10-year service life. The annual operating costs are \$2000 for each of the first two years, with an increase of \$600 per year thereafter. The MARR is 10%. When is the optimum time to retire the machine?

The cost curve for this type of problem is the net EUAC, evaluated at the MARR, as a function of time. From Section 7.4, we know that the net EUAC will be made up of two components: the capital recovery cost,

$$\mathrm{CR}(j) = (\$10\,000 - \$500)\,(A/P, 10\%, j) + (0.10)\,(\$500) \qquad (1)$$

and the equivalent annualized operating cost,

$$A(j) = \$1400 + \$600(A/G, 10\%, j) + \$600(P/F, 10\%, 1)\,(A/P, 10\%, j) \qquad (2)$$

In deriving (*1*), the constancy of the salvage value was used; in deriving (*2*), the gradient series was extended backwards by writing the first year's cost as \$1400 + \$600. In both expressions, j is the time, in years; thus, to keep the machine for 4 years would cost the company $\mathrm{CR}(4) + A(4)$ *per year.*

Substituting $j = 1, 2, \ldots, 10$ in (*1*) and (*2*), we generate Table 10-1; the points are plotted in Fig. 10-1. The data show that the machine should be retired at the end of 7 years.

The time interval for which the EUAC of an asset is smallest (7 years, for the machine of Example 10.1) is called the *economic life* of the asset. As we have seen, the economic life is given analytically as that value j^* at which

$$\mathrm{EUAC}(j) = \mathrm{CR}(j) + A(j)$$

is minimized.

The concept of economic life may also furnish the basis for a replacement decision, particularly when service lives are not precisely known, when salvage values in each year are not known, or when the equipment obsolesces rapidly.

Table 10-1

Years of Service Life, j	CR(j)	$A(j)$	EUAC(j) = CR(j) + $A(j)$
1	$10 500.00	$2 000.00	$12 500.00
2	5 523.81	2000.00	7 523.81
3	3 870.05	2181.24	6 051.29
4	3 046.97	2400.77	5 447.74
5	2 556.10	2629.99	5 186.09
6	2 231.30	2859.40	5 090.70
7	2 001.40	3085.07	5 086.47 = min.
8	1 830.68	3304.85	5 135.53
9	1 699.58	3518.12	5 217.70
10	1 596.13	3724.16	5 320.29

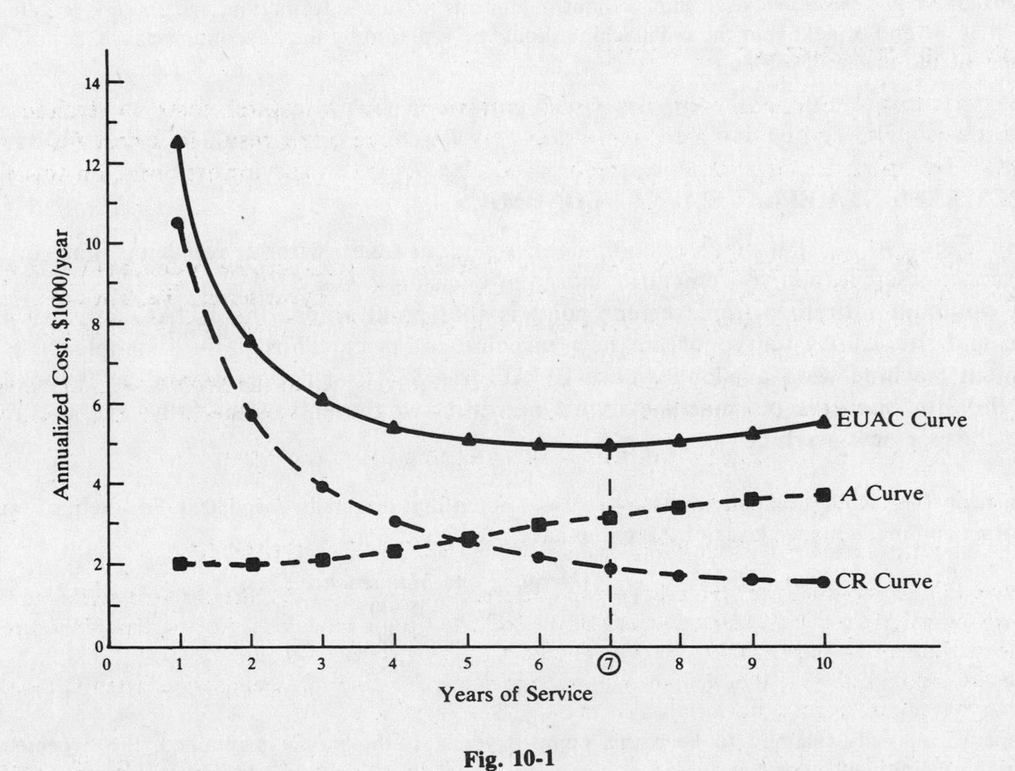

Fig. 10-1

Example 10.2 The XYZ Company purchased a very specialized machine three years ago for $25 000. This machine is not readily salable and is assumed to have a zero salvage value. Operating costs are expected to be $10 000 next year, and to increase by $800 per year thereafter. The company has an opportunity to replace the existing machine with another specialized one that will cost $12 000. This machine has no salvage value, a useful life of 10 years, and operating costs of $5000 in the first year, with an annual increase of $1200 thereafter. If the MARR is 15%, should the company replace the old machine with the new one?

For the new machine:

Year, j	$\text{EUAC}(j) = \$12\,000(A/P, 15\%, j) + \$5000 + \$1200(A/G, 15\%, j)$
1	\$18 800.00
2	12 940.52
3	11 344.16
4	10 793.96
$5 = j^*$	10 647.20
6	10 687.52
...	*(increases)*

The economic life of the new machine is thus 5 years, with a corresponding EUAC of \$10 647.20.

For the old machine,

$$\text{EUAC}(j) = A(j) = \$10\,000 + \$800(A/G, 15\%, j)$$

which is a strictly increasing function of j. Hence, $j^* = 1$ year, with a corresponding EUAC of \$10 000.

The old machine should be kept for its economic life of one more year, since its EUAC of \$10 000 is less than the EUAC of \$10 647.20 for the new machine. At the end of that year, the analysis should be repeated and updated to take any new information into account. If there is no new information, and the above data are still valid, then at the end of next year the old machine should be replaced by the new one, because, *at that time*, the EUAC for the old machine will be

$$\$10\,000 + \$800 = \$10\,800 > \$10\,647.20$$

10.3 RETIREMENT/REPLACEMENT ECONOMICS

The decision to retire a piece of equipment is seldom taken without replacing that equipment. Thus, in most cases, a joint retirement/replacement decision is made.

The optimum retirement/replacement point is that point where the EUAC curve of the old machine and the EUAC curve of the new machine intersect. Thus, if in Example 10.1 a new replacement machine were available whose EUAC was \$5500 at five years and \$4000 at all years beyond the fifth, then the old machine should be retired at the end of the fifth year, and replaced with the cheaper new machine.

Example 10.3 The XYZ Company owns a 4-year-old pump that originally cost \$3000. For the past four years the operating and maintenance costs of this pump have been:

Year of Service, k	*Operating and Maintenance Costs, C_k*
1	\$ 90
2	180
3	560
4	950

The company originally planned to keep this pump 8 years. If the pump is retained, the expected future operating and maintenance costs will be:

Year of Service, k	*Operating and Maintenance Costs, C_k*
5	\$1125
6	1500
7	1700
8	2000

The pump could be sold today as a used pump for $SV_0 = \$1200$. It is expected that the pump could be sold a year from now for \$900; two years from now for \$800; and afterwards for \$500. A new, energy-saving pump, with expected service life of 8 years, has just become available for $P = \$4000$. Its costs are as follows:

Year of Service, j	*Operating and Maintenance Costs, C'_j*
1	$ 40
2	80
3	260
4	450
5	625
6	1000
7	1200
8	1500

It is estimated that if this new pump is purchased it could be sold one year later for $3100, two years later for $2000, three years later for $1500, four years later for $1000, and thereafter for $900. Given a MARR of 15%, should the XYZ Company replace the old machine now, or at some later time?

In computing the EUAC curve of the old pump, the sunk costs of years 1 through 4 are disregarded. Thus, $k = 5$ becomes $j = 1$, and we have:

$$\text{EUAC}(j) = \text{CR}(j) + A(j) \qquad (j = 1, \ldots, 4) \tag{1}$$

where

$$\text{CR}(j) = (\text{SV}_0 - \text{SV}_j)(A/P, 15\%, j) + (0.15)\,\text{SV}_j$$
$$A(j) = [C_1(P/F, 15\%, 1) + \cdots + C_j(P/F, 15\%, j)](A/P, 15\%, j)$$

Similarly, for the new pump,

$$\text{EUAC}'(j) = \text{CR}'(j) + A'(j) \qquad (j = 1, \ldots, 8) \tag{2}$$

where

$$\text{CR}'(j) = (P - \text{SV}'_j)(A/P, 15\%, j) + (0.15)\,\text{SV}'_j$$
$$A'(j) = [C'_1(P/F, 15\%, 1) + \cdots + C'_j(P/F, 15\%, j)](A/P, 15\%, j)$$

Evaluating (*1*) for $j = 1, \ldots, 4$ and (*2*) for $j = 1, \ldots, 8$, we generate Table 10-2. It is seen that all the EUAC-values for the new pump are below the lowest EUAC-value for the old pump. Hence, the old pump should immediately be replaced by the new one. Even in the worst case, where the new pump is kept for only two years, its EUAC would still be lower than that for keeping the old pump for even one more year. The reader should note the oscillation of the EUAC curve for the new pump; the curve is *not* of the simple form shown in Fig. 10-1.

Table 10-2

Years From Now, j	EUAC(j)	EUAC′(j)
1	$1605.00	$1540.00
2	1663.90	1588.84
3	1796.30	1436.55
4	1852.18	1384.18
5		1402.41
6		1371.86
7		1368.52
8		1387.94

The question of how long the new pump should be kept is readily answered from its EUAC data: its economic life is 7 years. Of course, the unexpected advent of some new technology could alter these plans, just as the availability of the new pump altered the XYZ Company's original plans.

Comparative Use Value

It is sometimes useful to compare EUAC's via the *comparative use value* (CUV) of the current equipment relative to the replacement equipment. Let subscripts 1 and 2 pertain to the current and replacement machines, respectively; after n_1 and n_2 years of service, their annualized operating and maintenance costs are $A_1(n_1)$ and $A_2(n_2)$, and their salvage values are SV_1 and SV_2. Then, the CUV is given by

$$(\text{CUV} - SV_1)(A/P, i\%, n_1) + i\, SV_1 + A_1(n_1)$$
$$= (P_2 - SV_2)(A/P, i\%, n_2) + i\, SV_2 + A_2(n_2) \qquad (10.1)$$

where P_2 is the initial cost of machine 2. By (*10.1*), the CUV is the putative market price of machine 1 that makes $\text{EUAC}_1(n_1) = \text{EUAC}_2(n_2)$. If the actual market or current salvage value, P_1, is known, (*10.1*) can be solved algebraically to yield

$$\text{CUV} = P_1 + [\text{EUAC}_2(n_2) - \text{EUAC}_1(n_1)]\,(P/A, i\%, n_1) \qquad (10.2)$$

That is, the CUV of the current machine with respect to a replacement machine is the market value of the current machine plus the present value of all annual cost savings realized over the service period of the current machine.

It is evident from (*10.2*) that $\text{CUV} > P_1$ if and only if $\text{EUAC}_2(n_2) > \text{EUAC}_1(n_1)$; in this case, and only then, the current machine should be kept.

Example 10.4 Compute the CUV for keeping the old pump of Example 10.3 another four years, as against buying the new pump and keeping it seven years.

Substituting the numerical values from Example 10.3 and Appendix A into (*10.2*), we find

$$\text{CUV} = \$1200 + (\$1368.52 - \$1852.18)(0.35027)^{-1}$$
$$= \$1200 - \$1381 = -\$181$$

The CUV is less than the current salvage value by \$1381, which amount represents the present-worth *loss* that would be incurred in keeping the old pump four more years.

Present-Worth Method for Equal Service Periods

If machines 1 and 2 have present-worth *costs* PW_1 and PW_2, and service periods $n_1 = n_2 = n$, then

$$PW_1 = \text{EUAC}_1(n)\,(P/A, i\%, n) \qquad PW_2 = \text{EUAC}_2(n)\,(P/A, i\%, n)$$

and so

$$PW_1 - PW_2 = [\text{EUAC}_1(n) - \text{EUAC}_2(n)]\,(P/A, i\%, n)$$

Thus, a comparison of EUAC's may be replaced by a comparison of PW's (although usually there is no computational advantage in so doing).

Short-Study-Period Method for Unequal Service Periods

In the case $n_1 \neq n_2$, we might compare the costs of a series of machines 1 and a series of machines 2 over the least common multiple of n_1 and n_2; this approach will be illustrated in Section 10.4. However, if we do not want to assume a continual substitution of machines in kind, we may restrict the study period to the smallest of n_1, n_2, and the forecasting horizon. Thus, only known data are included and tenuous estimates are ruled out. In this short-study-period approach, any "unused" values or costs are distributed back over the study period.

Example 10.5 The ABC Company is contemplating replacing its current milling machine with an improved machine. Data are as shown in Table 10-3. The executives at the ABC Company do not feel that any estimates beyond 10 years are accurate for decision making purposes. If the MARR is 15%, should the company replace the current machine with the new improved one?

The length of the study period is the minimum of 15, 10, and 10 years; i.e., 10 years. The present-worth *cost* of 15 more years of service from the current machine is

$$PW_1 = \$500 - \$100(P/F, 15\%, 15) + \$1000(P/A, 15\%, 15)$$
$$= \$500 - \$100(8.1371)^{-1} + \$1000(0.17102)^{-1} = \$6334.99$$

Table 10-3

	Salvage Value				
	Now	End of Life	Original Cost	Annual Cost	Service Life
Current Machine	$500	$100	$5000	$1000	15 more years
Improved Machine	—	500	7000	375	10 years

Annualizing this over the 10-year study period gives

$$\text{EUAC}_1(10) = (\$6334.99)(A/P, 15\%, 10) = (\$6334.99)(0.19925) = \$1262.25$$

For the improved machine:

$$\begin{aligned}\text{EUAC}_2(10) &= (\$7000 - \$500)(A/P, 15\%, 10) + (0.15)(\$500) + \$375 \\ &= \$6500(0.19925) + \$75 + \$375 = \$1745.13\end{aligned}$$

Thus, keeping the current machine is $\$1745.13 - \$1262.25 = \$482.88$ cheaper (per year for 10 years) than replacing it with the improved machine.

Example 10.6 Suppose that in Example 10.5 the company executives feel they cannot accurately forecast annual costs beyond five years into the future. All other data remain the same. Using a five-year study period, we obtain for the current machine:

$$\text{EUAC}_1(5) = (\$6334.99)(A/P, 15\%, 5) = \$1889.85$$

and for the improved machine:

$$\text{EUAC}_2(5) = (\$7000 - \$500)(A/P, 15\%, 5) + (0.15)(\$500) + \$375 = \$2389.08$$

The numbers are different, but the decision is the same as before: it is cheaper to keep the current machine. The difference between the alternatives in the 5-year study is $\$2389.08 - \$1889.85 = \$499.23$, which is slightly larger than the \$482.88 difference found in the 10-year study of Example 10.5. In this case, using the shorter study period did not reduce the amount of discrimination between the alternatives.

10.4 REPLACEMENT ASSUMPTION FOR UNEQUAL-LIVED ASSETS

Equipment investment decisions frequently involve the comparison of assets with unequal lives. In most businesses, a piece of equipment will be replaced with a like one at the end of its useful life, in order for the firm to continue operating. When this is the case, the cash flows of the unequal-lived assets should be estimated into the future until the least common multiple of their individual useful lives has been reached. The EUAS method can then be readily applied.

Example 10.7 A company can purchase either of two alternative machines, A and B, with the following characteristics:

	P	A	n
Machine A	$15 000	$6000	5
Machine B	9 000	3000	3

Here, P is the initial outlay, A is the annual net cash flow, and n is the useful life of the machine. It is assumed that when either machine is at the end of its life, a similar replacement machine will be purchased. Which machine should be purchased, under a MARR of 10%?

Machines A and B will first reach a common multiple of their individual useful lives at the end of fifteen years. The cash flows for this fifteen-year period are displayed in Table 10-4. Note the replacements of machine A by exact replicas at the end of five and ten years; and of machine B, at the end of three, six, nine, and twelve years. (Both machines are also replaced at the end of fifteen years, but those costs belong to the next study period.)

Table 10-4

End of Year	Machine A	Machine B
0	−$15 000	−$9000
1	6 000	3000
2	6 000	3000
3	6 000	3000 − 9000
4	6 000	3000
5	6 000 − 15 000	3000
6	6 000	3000 − 9000
7	6 000	3000
8	6 000	3000
9	6 000	3000 − 9000
10	6 000 − 15 000	3000
11	6 000	3000
12	6 000	3000 − 9000
13	6 000	3000
14	6 000	3000
15	6 000	3000

Proceeding as in Example 7.4, we compute:

$$\text{EUAS}_A = -\$15\,000(A/P, 10\%, 5) + \$6000$$
$$= \$2043 \text{ for fifteen years}$$
$$\text{EUAS}_B = -\$9000(A/P, 10\%, 3) + \$3000$$
$$= -\$619 \text{ for fifteen years}$$

The conclusion is that machine B is unacceptable, and that machine A should be purchased.

Solved Problems

10.1 Mr. Jones bought a new car in September 1981 for $7800. He paid $2400 down, and financed the balance with a loan at 18% nominal interest to be repaid in 35 monthly installments of $199.42 each. It is understood that if Mr. Jones fails to make a payment, the car will be repossessed by the loan company and, though Mr. Jones will owe nothing, he will lose all money already paid. He can also pay the outstanding balance of the loan at any time. Twelve months after the transaction, Mr. Jones's balance is $3855. Thanks to a good deal, Mr. Jones has this amount and is going to pay off the loan. At this point, Mr. Jones's brother offers him an essentially identical car (they bought the same model the same day, and the use and maintenance of both cars have been very similar) for $3500. Mrs. Jones would prefer to keep their current car "because we have already spent almost $4800 on it." What is the sensible decision to make?

This is a clear and very common instance of a sunk cost: Mrs. Jones is wrong in her analysis. While it is true that they have invested a large amount of money so far, it is also true that this money will not be recovered, no matter what decision is taken. The sole question is whether Mr. Jones should acquire a car for \$3855 or acquire a car for \$3500. Provided the two cars are physically equivalent, the answer is obvious.

10.2 A machine can be sold now for \$15 000; if kept for another year, its salvage value will decline to \$13 000. The operating expenses for this year are expected to be \$30 000. A new machine is available for \$50 000, with expected operating expenses of \$18 000 for the first year, increasing by \$1000 a year because of deterioration. It is believed that after 5 years new technology would make replacement necessary; the new machine's salvage value at that time is estimated to be \$20 000. The MARR is 20%. Should the new machine be acquired?

By (*10.1*),

$$(\text{CUV} - \$13\,000)\,(A/P, 20\%, 1) + (0.20)\,(\$13\,000) + \$30\,000$$
$$= (\$50\,000 - \$20\,000)\,(A/P, 20\%, 5) + (0.20)\,(\$20\,000) + \$18\,000 + \$1000(A/G, 20\%, 5)$$

whence CUV = \$13 893.25. Since the comparative use value of the old machine is smaller than its net market value (\$15 000), the machine should be replaced.

10.3 A contractor can purchase a used machine for \$1000. The market value of the machine is expected to decrease \$70 the first year and \$60 per year, the second and third years. Operating disbursement is estimated at \$8000 the first year and is expected to increase by \$175 a year thereafter. An alternative is to buy a new machine costing \$10 000. It is believed that the salvage value of this machine will decrease by 15% each year over a maximum service life of 20 years. The operating expenses are estimated at \$6050 the first year and are expected to increase by \$135 a year after that. If the MARR is 10%, which machine should be bought, and when should that machine be replaced? (Assume that the used machine is unique, but that new machines are always available.)

For the used machine:

$$\text{EUAC}_1(1) = (\$1000 - \$930)\,(A/P, 10\%, 1) + (0.10)\,(\$930) + \$8000 = \$8170.00$$
$$\text{EUAC}_1(2) = (\$1000 - \$870)\,(A/P, 10\%, 2) + (0.10)\,(\$870) + \$8000 + \$175(A/G, 10\%, 2)$$
$$= \$8245.24$$
$$\text{EUAC}_1(3) = (\$1000 - \$810)\,(A/P, 10\%, 3) + (0.10)\,(\$810) + \$8000 + \$175(A/G, 10\%, 3)$$
$$= \$8321.30$$

For the new machine:

$$\text{EUAC}_2(1) = [\$10\,000 - (0.85)\,(\$10\,000)]\,(A/P, 10\%, 1) + (0.10)\,(0.85)\,(\$10\,000) + \$6050$$
$$= \$8550.00$$
$$\text{EUAC}_2(2) = [\$10\,000 - (0.85)^2(\$10\,000)]\,(A/P, 10\%, 2) + (0.10)\,(0.85)^2(\$10\,000)$$
$$+ \$6050 + \$135(A/G, 10\%, 2) = \$8435.71$$
$$\text{EUAC}_2(3) = \$8342.20$$
$$\text{EUAC}_2(4) = \$8206.37$$
$$\text{EUAC}_2(5) = \$8435.11$$

The used machine should be bought and kept for one year (its economic life). At that time, the new machine should be bought and kept for four years (its economic life).

10.4 A machine costs \$10 000 and is expected to have scrap value \$1500 whenever it is retired. The operating disbursements for the first year are expected to be \$1500 and they will then increase \$400 per year, as a result of deterioration. If the MARR is 15%, determine the machine's economic life.

For $j = 1, 2, \ldots,$

$$\text{CR}(j) = (\$10\,000 - \$1500)(A/P, 15\%, j) + (0.15)(\$1500)$$
$$A(j) = \$1500 + \$400(A/G, 15\%, j)$$
$$\text{EUAC}(j) = \text{CR}(j) + A(j)$$

The evaluations, Table 10-5, give an economic life of 8 years.

Table 10-5

Years of Service, j	CR(j)	$A(j)$	EUAC(j)
1	\$10 000.00	\$1500.00	\$11 500.00
2	5 453.52	1686.05	7 139.57
3	3 947.83	1862.85	5 810.68
4	3 207.30	2030.52	5 232.82
5	2 760.72	2189.12	4 949.84
6	2 471.04	2338.88	4 809.92
7	2 268.06	2479.96	4 748.02
8	2 119.23	2612.52	4 731.75
9	2 006.35	2736.88	4 743.23
10	1 918.63	2853.28	4 771.91

10.5 A plant is considering buying a second-hand machine to use as stand-by equipment. The machine costs \$3000 and has an economic life of 10 years, at which time its salvage value is \$600; expected annual operating costs are \$100. Without a stand-by machine, the plant would have to shut down an average of seven days a year at a cost of \$50 per day. If the MARR is 10%, is it expedient to buy the stand-by machine?

$$\text{EUAC}_{\text{shut-down}} = (7)(\$50) = \$350$$
$$\text{EUAC}_{\text{stand-by}} = (\$3000 - \$600)(A/P, 10\%, 10) + (0.10)(\$600) + \$100 = \$550$$

The stand-by machine should not be purchased.

10.6 XYZ Company is considering replacing a machine. The new improved machine will cost \$16 000 installed; it will have an estimated service life of 8 years and \$3000 salvage value. It is estimated that operating expenses will average \$1000 a year. The present machine was purchased for \$20 000 four years ago and is estimated to have 8 more years of service life, at the end of which its salvage value will be \$2000. Operating costs are \$1800 per year. If replaced now, it can presumably be sold for \$5000. Using a MARR of 15%, determine whether to replace the existing machine.

Compare present-worth costs over the next 8 years.

$$\text{PW}_{\text{new}} = \$16\,000 + \$1000(P/A, 15\%, 8) - \$3000(P/F, 15\%, 8) = \$19\,507$$
$$\text{PW}_{\text{existing}} = \$5000 + \$1800(P/A, 15\%, 8) - \$2000(P/F, 15\%, 8) = \$12\,423$$

Do not replace the machine.

10.7 Consider the replacement situation indicated in Table 10-6. If estimates beyond 8 years are unreliable and if the MARR is 15%, decide whether it is expedient to replace the current equipment.

The study period is limited by the current equipment and the forecast horizon to 8 years.

$$\text{EUAC}_{\text{current}}(8) = (\$8000 - \$1000)(A/P, 15\%, 8) + (0.15)(\$1000) + \$3600 = \$5310$$

Table 10-6

	Salvage Value		Original Cost	Annual Cost	Service Life
	Now	End of Life			
Current Equipment	$8000	$1000	$17 000	$3600	8 more years
Replacement Candidate		6000	19 000	800	15 years

For the replacement candidate, the present-worth cost is given by

$$\text{PW}_{\text{replacement}} = \$19\,000 + \$800(P/A, 15\%, 15) - \$6000(P/F, 15\%, 15) = \$22\,940$$

Annualizing over the 8-year period,

$$\text{EUAC}_{\text{replacement}}(8) = \$22\,940(A/P, 15\%, 8) = \$5112$$

Replace the current equipment.

10.8 To keep an existing machine going for a number of years, an extensive (and expensive: $4000) overhaul is needed. Maintenance is expected to be $2000 annually for the next 2 years and to increase by $1000 per year after that. The machine has no present or future salvage value. An alternative machine costs $8000 and, owing to its specialized nature, it has no salvage value after it is installed. Maintenance expenses are expected to be $1000 the first year, increasing by $500 per year in subsequent years. If the MARR is 15%, determine the best course of action.

For the defender, writing the first year's maintenance as $1000 + $1000,

$$\text{EUAC}_1(j) = \$4000(A/P, 15\%, j) + \$1000(P/F, 15\%, 1)\,(A/P, 15\%, j) + \$1000 + \$1000(A/G, 15\%, j)$$

and for the challenger,

$$\text{EUAC}_2(j) = \$8000(A/P, 15\%, j) + \$1000 + \$500(A/G, 15\%, j)$$

From the evaluations in Table 10-7, we see that the economic lives are $j_1^* = 4$ years and $j_2^* = 7$ years. It follows that the best course of action is to overhaul the current equipment and keep it for four years. Then (provided an analysis indicates that things remain as expected), buy the new equipment and keep it for seven years.

Table 10-7

j	$\text{EUAC}_1(j)$	$\text{EUAC}_2(j)$
1	$6600	$10 200
2	4460	6 154
3	4040	4 957
4	4032	4 465
5	4175	4 248
6		4 163
7		4 148
8		4 173

10.9 It is necessary to pump twice as much water as can be handled by the existing small pump, which is now 5 years old. This pump can be sold now for $1200 or kept for 5 years, after which it will have zero salvage value. Operating expenses are $3000 per year. If the pump is kept, a similar one must be purchased for $3500, with operating costs of $2500 per year; its salvage values after 5 and 10 years are the same as for the original pump. A large pump, equal in capacity to the two small pumps, costs $6000, with operating expenses of $4500 per year. New machines have economic lives of 10 years, and zero salvage values at that date. Analyze the situation, if the MARR is 10%.

There are 3 possible alternatives: (1) to replace the present pump by a new, large pump; (2) to buy a new small pump now and every fifth year hereafter; (3) to buy a new small pump now, and after 5 years to sell it and install a single large pump. We calculate the present-worth costs for a 10-year study period.

plan 1 $PW_1 = \$6000 + \$4500(P/A, 10\%, 10) = \$33\,651$

plan 2 $PW_2 = \$1200 + \$3000(P/A, 10\%, 5) + \$3500 + \$2500(P/A, 10\%, 10)$
$+ [\$3500(A/P, 10\%, 10) + \$2500]\,(P/A, 10\%, 5)\,(P/F, 10\%, 5)$
$= \$38\,658$

plan 3 $PW_3 = \$1200 + \$3000(P/A, 10\%, 5) + \$3500 + \$2500(P/A, 10\%, 5) - \$1200(P/F, 10\%, 5)$
$+ [\$6000(A/P, 10\%, 10) + \$4500]\,(P/A, 10\%, 5)\,(P/F, 10\%, 5)$
$= \$37\,694$

Note, in the calculations for plans 2 and 3, how certain costs are spread out over ten years and then are partially brought back into the study period.

The conclusion is that plan 1 is best, with plan 3 being slightly superior to plan 2.

10.10 A company decides to automate a process by installing a machine that costs \$8000 and is expected to save \$2500 per year. Discuss the wisdom of this decision, if the economic life of the machine is 10 years, at which time it has a \$2000 salvage value, and if the MARR is 15%.

The present worth of the decision is (costs counted negative):

$$PW = -\$8000 + \$2000(P/F, 15\%, 10) + \$2500(P/A, 15\%, 10) = +\$5041.38$$

that is, the company can expect net savings of \$5041.38 (today's dollars) over the next 10 years.

Supplementary Problems

10.11 Solve Problem 10.10 for an economic life of 3 years.
Ans. PW = −\$976.94 (the company should not automate)

10.12 Table 10-8 shows the expected annual operating costs and salvage values for a machine whose initial cost is \$20 000. Find the economic life of the machine, if the MARR is 20%. *Ans.* 6 years

Table 10-8

Year of Service	Salvage Value at End of Year	Operating Cost for Year
1	\$10 000	\$ 2 000
2	9 000	3 000
3	8 000	4 000
4	7 000	5 000
5	6 000	6 000
6	5 000	7 000
7	4 000	8 000
8	3 000	9 000
9	2 000	10 000
10	1 000	11 000

10.13 For a MARR of 15%, find the economic life of the new machine in Problem 10.3. *Ans.* 15 years

10.14 Repeat Problem 10.7 for a forecast horizon of 5 years.
Ans. $\text{EUAC}_{\text{current}}(5) = \$5838$, $\text{EUAC}_{\text{replacement}}(5) = \$5843$; therefore, keep the current equipment. (Note that this is a conservative criterion, since it tends to preserve the existing situation when the challenger's edge consists in savings in the distant future, the estimation of which must be unreliable.)

10.15 For the situation described in Problem 10.5, determine the number of down-days per year that would justify the acquisition of the stand-by machine. *Ans.* 12.01 (thus, 13)

10.16 A word-processor was bought two years ago for $22 000. At the time, the machine was expected to last six years and to have operating costs of $7200 the first year, increasing by $300 per year thereafter. The salvage value at the end of the sixth year is assumed to be zero. Another company is presently offering a competitive machine for $16 000; they will give $10 000 for the one in use as trade-in value. Although the book value of the old machine is $14 167, this offer of $10 000 is thought to be fair, since the technical obsolescence of the current system would make it very difficult to get a better offer. For this reason, it is believed that the salvage value of the current machine will decrease by $2500 per year over the next four years. The new machine is expected to last four years and to have operating expenses of $6500 per year. After four years, its salvage value will be zero. If the MARR is 15%, should the machine be changed? If so, when?
Ans. The old machine's current market value (salvage value) is $10 000; its economic life is found as 4 more years, with $\text{EUAC}_{\text{old}}(4) = \$11\,700.59$. The new machine also has an economic life of 4 years, with $\text{EUAC}_{\text{new}}(4) = \$12\,104.32$. Thus, the old machine should be kept for four more years.

10.17 Refer to Table 10-9. If the MARR is 15%, should the current machine be replaced?
Ans. No (based on a 10-year study period)

Table 10-9

	Salvage Value		Original Cost	Annual Cost	Service Life
	Now	End of Life			
Current Machine	$14 000	$1000	$50 000	$10 000	10 more years
Improved Machine		5000	45 000	3 750	15 years

10.18 A pump costing $18 000 is expected to have operating disbursements of $6500 the first year. The machine's resale value is expected to decline by 15% a year, while its operating expenses are expected to increase by $500 a year. If the MARR is 20%, determine the economic life and the corresponding annual equivalent cost. *Ans.* 8 years, EUAC(8) = $12 181.50

10.19 In a replacement analysis, data for the challenger are as follows:

Initial cost (installed): $12 000

Maximum service life: 8 years

Operating expenses: none, the first three years; $2000 the fourth and fifth years; increasing by $2500 per year after the fifth year

Salvage value: zero at all times

The MARR is 10%. Tabulate the EUAC of the challenger and infer its economic life.
Ans. See Table 10-10.

10.20 Is it expedient to replace the current machine of Table 10-11, if the MARR for this type of study is 15%?
Ans. No: CUV = $1088 > $1000

Table 10-10

j	EUAC(j)
1	$13 200
2	6 914
3	4 825
4	4 390
5	4 071
$6 = j^*$	3 706
7	3 843
8	4 162

Table 10-11

	Current Machine	Challenger
Service Life	3 more years	7 years
Original Cost		$5500
Salvage Value	$1000 now, $200 after 3 years	$500 after 7 years
Annual Cost	$1300	$600

10.21 Compute the CUV of the new machine in Problem 10.13, if the old machine is kept two years and the new one is kept ten years. *Ans.* CUV = $9448 (<$10 000).

10.22 The data in Table 10-12 pertain to the average standard-size 1979-model automobile, purchased for $6263. Assuming that 12% is a good estimate of the pertinent interest rate, determine the economic life of such an automobile. *Ans.* 10 years

Table 10-12

Years of Service	Operating Cost	Salvage Value
1	$2178	$4503
2	1880	3582
3	2080	2881
4	1554	2255
5	2166	1723
6	1923	1285
7	2379	890
8	1476	564
9	1604	251
10	1078	0

10.23 A 4-year-old die-casting machine, of market value $3500, is 50% too small for future production needs. A new machine with identical production capacity costs $5000 installed. Both machines are expected to have economic lives of 6 years from this date. Salvage values at that date will be $1000 for the new, and $700 for the old, machine. Annual operating expenses for the new and old machines are expected to be $3500 and $4000, respectively. A double-capacity machine is also available; its installed cost is $12 000, with a salvage value of $2000 at the end of its 6-year economic life. Operating costs are expected to be $6000 per year. If the MARR is 10%, which machine should be purchased?
Ans. $\text{EUAC}_{\text{new,small}}$ = $9231, $\text{EUAC}_{\text{large}}$ = $8496; buy the large machine.

10.24 A certain 5-year-old machine has a salvage value of $1200 if sold today, and of $400 if sold 5 years from now. Its operating expenses are $800 per year. A new improved machine is available for $2400, has expected operating expenses of $500 per year, and has a salvage value of $1000 at the end of its 5-year economic life. There is also the possibility of overhauling the old machine at a cost of $600, which would increase the salvage value in 5 years by $200 and reduce the operating expenses by $200 per year. If the MARR is 10%, which course of action should be taken?
Ans. EUAC_{old} = $1051, EUAC_{new} = $969, $\text{EUAC}_{\text{overhaul}}$ = $977; buy the new machine.

Chapter 11

Depreciation and Taxes

11.1 DEFINITIONS

Depreciation is a way of accounting for the cost of an asset when income is determined for tax purposes. The cost, including any delivery or installation charges, is treated as a prepayment for future services; and depreciation consists in amortizing this prepayment over the period of use of the asset.

The *annual depreciation* is the amount of the asset's cost that is charged off in a given year; the total of the annual depreciations to date is the *accumulated depreciation*. The *salvage value* or *scrap value* of an asset is the estimated proceeds that will be realized from its sale or disposition when it is retired. Under federal tax law, the *net salvage value* is either zero or the salvage value minus the cost of removing the asset from the premises, whichever is greater. The *adjusted cost* of an asset is its original cost less its net salvage value.

The *useful life*, over which an asset is depreciated, may not be the same as its service life, physical life, economic life, market life, etc. The U.S. government publishes guidelines for most equipment, showing the ranges of useful lives allowed in tax computations.

Depreciable And Nondepreciable Assets

The current tax laws permit only assets with a useful life or more than one year to be depreciated. Depreciation is allowed only for assets used in a business, trade or profession, or held for the production of income. Personal property, such as a family residence or automobile used for pleasure, is not depreciable; however, that portion of an automobile or other property which is used in business may be depreciable. Depreciation is not permitted on land (or on its upkeep), even when it is used for business purposes or income generation. However, buildings and equipment which occupy that land are depreciable if used in a business or to generate income. Inventories of goods used in a business, other stock in trade, and short-term assets that will be consumed during a normal year's operation of the business are not depreciable.

Computation Methods

The four traditional methods of computing an asset's depreciation from its cost, useful life, and salvage value will be presented in Sections 11.2–11.5. A newer method is treated in Section 11.11.

11.2 STRAIGHT-LINE METHOD

For an asset with useful life n years, the annual depreciation in year j is

$$\mathrm{SD} \equiv \frac{\text{adjusted cost}}{n} \qquad (j = 1, 2, \ldots, n) \tag{11.1}$$

a constant independent of j; this corresponds to the constant annual depreciation *rate* $r_s \equiv 100\%/n$. The accumulated depreciation at the end of year j is simply

$$\mathrm{ASD}_j \equiv j \times \mathrm{SD} \tag{11.2}$$

and the *book value* of the asset at the end of year j is defined as

$$\mathrm{SB}_j \equiv (\text{original cost}) - \mathrm{ASD}_j \tag{11.3}$$

In particular, for $j = n$, (*11.3*) gives

$$\mathrm{SB}_n = (\text{original cost}) - (\text{adjusted cost}) = \text{net salvage value}$$

Example 11.1 A new machine costs \$160 000, has a useful life of 10 years, and can be sold for \$15 000 at the end of its useful life. It is expected that \$5000 will be spent to dismantle and remove the machine at the end of its useful life. Determine the straight-line depreciation schedule for this machine.

Here, the adjusted cost is \$160 000 − (\$15 000 − \$5000) = \$150 000, and the rate of depreciation is 10% per year. Applying (*11.1*), (*11.2*), and (*11.3*), we generate Table 11-1. At the end of year 10, the sale of the asset for \$15 000 will remove the \$10 000 book value from the firm's accounting records.

Table 11-1

Year, j	Depreciation Charge (10%) for Year, SD	Accumulated Depreciation, ASD_j	Book Value at End of Year, SB_j
1	\$15 000	\$ 15 000	\$145 000
2	15 000	30 000	130 000
3	15 000	45 000	115 000
4	15 000	60 000	100 000
5	15 000	75 000	85 000
6	15 000	90 000	70 000
7	15 000	105 000	55 000
8	15 000	120 000	40 000
9	15 000	135 000	25 000
10	15 000	150 000	10 000

11.3 DECLINING-BALANCE METHOD

In this method, the annual depreciation in year j is computed as a fixed fraction of the asset's book value at the end of year $j-1$:

$$\mathrm{DD}_j = \frac{r_d}{100\%} \times \mathrm{DB}_{j-1} \qquad (j = 1, 2, \ldots, n) \tag{11.4}$$

where r_d is the declining-balance annual depreciation rate, in percent, and $\mathrm{DB}_0 \equiv$ original cost. The accumulated depreciation at the end of year j is

$$\mathrm{ADD}_j \equiv \sum_{k=1}^{j} \mathrm{DD}_k \tag{11.5}$$

and the book value at the end of year j is

$$\mathrm{DB}_j \equiv \mathrm{DB}_0 - \mathrm{ADD}_j \tag{11.6}$$

Equations (*11.4*), (*11.5*), and (*11.6*) imply the recursion formula

$$\mathrm{DB}_j = \left(1 - \frac{r_d}{100\%}\right)\mathrm{DB}_{j-1} \qquad (j = 1, 2, \ldots, n)$$

which may be solved to give the following explicit expressions ($k = 1, 2, \ldots, n$):

$$\mathrm{DB}_k = \left(1 - \frac{r_d}{100\%}\right)^k \mathrm{DB}_0$$

$$\mathrm{DD}_k = \mathrm{DB}_{k-1} - \mathrm{DB}_k = \left(1 - \frac{r_d}{100\%}\right)^{k-1} \left(\frac{r_d}{100\%}\right) \mathrm{DB}_0 \tag{11.7}$$

$$\mathrm{ADD}_k = \mathrm{DB}_0 - \mathrm{DB}_k = \left[1 - \left(1 - \frac{r_d}{100\%}\right)^k\right] \mathrm{DB}_0$$

We see that the depreciation amount (and also the book value) decreases geometrically with time. Thus, the declining-balance method results in a larger share of the depreciation being charged during the earlier years of the asset's life. In contrast to the straight-line method, it is an *accelerated* depreciation method.

Also unlike the straight-line method, the declining-balance method does not *automatically* take account of the net salvage value of the asset. Thus, according to the first equation (*11.7*), the book value steadily decreases through positive values, and could very well become smaller than the net salvage value. Salvage value is included in the method by fiat: Federal tax law forbids the application of the method past the point at which ADD_k becomes greater than the adjusted cost of the asset—which is precisely the point at which DB_k becomes smaller than the net salvage value.

Example 11.2 Apply the *double-declining-balance* method (i.e., $r_d = 2r_s = 200\%/n$) to (*a*) the machine of Example 11.1; (*b*) the machine of Example 11.1, with the net salvage value changed to \$30 000.

In either case, construct the depreciation schedule by applying (*11.4*), with $r_d = 20\%$ and $\mathrm{DB}_0 = \$160\,000$, for as long as is permitted.

(*a*) See Table 11-2. Here, the accumulated depreciation never reaches \$150 000, the adjusted cost of the machine. The excess of the final book value over the net salvage value,

$$\$17\,179.87 - \$10\,000 = \$7179.87$$

will presumably be deducted from the firm's income as a capital loss, upon sale of the machine.

Table 11-2

Year, j	Depreciation Charge (20%) for Year, DD_j	Accumulated Depreciation, ADD_j	Book Value at End of Year, DB_j
1	\$32 000	\$ 32 000	\$128 000
2	25 600	57 600	102 400
3	20 480	78 080	81 920
4	16 384	94 464	65 536
5	13 107.20	107 571.20	52 428.80
6	10 485.76	118 056.96	41 943.04
7	8 388.61	126 445.57	33 554.43
8	6 710.88	133 156.45	26 843.55
9	5 368.71	138 525.16	21 474.84
10	4 294.96	142 820.12	17 179.87

(*b*) See Table 11-3. Entries for years 1 through 7 are computed in normal fashion. The depreciation charge for the 8th year becomes \$3554.43; any larger amount would cause the accumulated depreciation to exceed the legal maximum of

$$\$160\,000 - \$30\,000 = \$130\,000$$

No depreciation can be taken in years 9 and 10. A total of \$30 000 in book value, equal to the machine's net salvage value, remains undepreciated. The sale of the machine for its salvage value will remove this \$30 000 from the firm's accounting records.

Table 11-3

Year	Depreciation Charge for Year	Accumulated Depreciation	Book Value at End of Year
1	$ 32 000	$ 32 000	$128 000
2	25 600	57 600	102 400
3	20 480	78 080	81 920
4	16 384	94 464	65 536
5	13 107.20	107 571.20	52 428.80
6	10 485.76	118 056.96	41 943.04
7	8 388.61	126 445.57	33 554.43
8	3 554.43	130 000	30 000
9	0	130 000	30 000
10	0	130 000	30 000

11.4 SUM-OF-YEARS'-DIGITS METHOD

The sum of years, SY, for an asset with useful life n years is

$$\mathrm{SY} \equiv \sum_{j=1}^{n} j = 1 + 2 + \cdots + n = \frac{n(n+1)}{2} \tag{11.8}$$

In the sum-of-years'-digits method, the annual depreciation in year j is given by

$$\mathrm{SYD}_j = \frac{n+1-j}{\mathrm{SY}} \times (\text{adjusted cost}) \qquad (j = 1, 2, \ldots, n) \tag{11.9}$$

whence the accumulated depreciation at the end of year j is given by

$$\mathrm{ASYD}_j = \frac{jn - [j(j-1)/2]}{\mathrm{SY}} \times (\text{adjusted cost}) \tag{11.10}$$

The book value at the end of year j is defined in the usual way:

$$\mathrm{SYB}_j \equiv (\text{original cost}) - \mathrm{ASYD}_j \tag{11.11}$$

Because ASYD_n = adjusted cost, SYB_n = net salvage value, as in the straight-line method.

Table 11-4

Year, j	Depreciation Charge for Year, SYD_j	Accumulated Depreciation, ASYD_j	Book Value at End of Year, SYB_j
1	$27 272.73	$ 27 272,73	$132 727.27
2	24 545.45	51 818.18	108 181.82
3	21 818.18	73 636.36	86 363.64
4	19 090.91	92 727.27	67 272.73
5	16 363.64	109 090.91	50 909.09
6	13 636.36	122 727.27	37 272.73
7	10 909.09	133 636.36	26 363.64
8	8 181.82	141 818.18	18 181.82
9	5 454.55	147 272.73	12 727.27
10	2 727.27	150 000.00	10 000.00

According to (*11.8*), a different fraction is applied each year to the adjusted cost of the asset to obtain the annual depreciation. The denominator of this fraction is the total of the digits representing the years of the estimated useful life of the asset; the numerator changes each year, so as to represent the number of years of useful asset life remaining at the start of that year. Currently, the tax laws prohibit the sum-of-years'-digits method from being used on any property for which the double-declining-balance method is prohibited.

Example 11.3 Apply the sum-of-years'-digits method to the machine of Example 11.1.

By (*11.8*), SY = (10)(11)/2 = 55, and the adjusted cost of the machine is $150 000. Repeated application of (*11.9*) generates Table 11-4.

11.5 SINKING-FUND METHOD

This method depreciates an asset *as if* the firm were to make a series of equal annual deposits (a *sinking fund*) whose value at the end of the asset's useful life just equaled the cost of replacing the asset. Writing

$A' \equiv$ sinking-fund deposit

$C \equiv$ (purchase price of replacement asset) − (net salvage value of current asset)

$n \equiv$ useful life of current asset

$i \equiv$ annual interest rate

we have: $A' = C\,(A/F, i\%, n)$. The amount in the sinking fund at the end of year j ($j = 1, 2, \ldots, n$) is identified with the accumulated depreciation to date; thus,

$$\text{ASFD}_j \equiv A'\,(F/A, i\%, j) = C\,(A/F, i\%, n)\,(F/A, i\%, j) \tag{11.12}$$

and the depreciation amount in year j is

$$\begin{aligned} \text{SFD}_j &= \text{ASFD}_j - \text{ASFD}_{j-1} \\ &= C\,(A/F, i\%, n)\,[(F/A, i\%, j) - (F/A, i\%, j-1)] \\ &= C\,(A/F, i\%, n)\,(1+i)^{j-1} \end{aligned} \tag{11.13}$$

As usual, the book value is defined as

$$\text{SFB}_j \equiv (\text{original cost}) - \text{ASFD}_j \tag{11.14}$$

Table 11-5

Year, j	Depreciation Charge for Year, SFD_j	Accumulated Depreciation, ASFD_j	Book Value at End of Year, SFB_j
1	$ 7 387.81	$ 7 387.81	$152 612.19
2	8 495.98	15 883.79	144 116.21
3	9 770.38	25 654.17	134 345.83
4	11 235.93	36 890.10	123 109.90
5	12 921.32	49 811.43	110 188.57
6	14 859.52	64 670.95	95 329.05
7	17 088.45	81 759.40	78 240.60
8	19 651.72	101 411.12	58 588.88
9	22 599.48	124 010.60	35 989.40
10	25 989.40	150 000.00	10 000.00

It is seen from (*11.13*) that the annual depreciation amount *increases* geometrically with time—just the opposite of the declining-balance method. As a matter of tax law, the sinking-fund method may be used only when the replacement asset will have the same original cost as the current asset, in which case C = adjusted cost of current asset. (Otherwise, the firm could take a total depreciation allowance in excess of the current asset's adjusted cost, and this is not allowed.)

Example 11.4 Apply the sinking-fund method to the machine of Example 11.1, given $i = 15\%$.

By repeated application of (*11.13*), starting with

$$\text{SFD}_1 = \$150\,000(A/F, 15\%, 10) = \$7387.81$$

Table 11-5 is generated.

11.6 GROUP AND COMPOSITE DEPRECIATION

The methods discussed in Sections 11.2–11.5 are all *unit depreciation methods* in that they apply to a single item, asset, or unit. When there are many like items, a *group depreciation* becomes convenient, whereby a single annual depreciation figure is computed for the ensemble, using the sum of the items' original costs and the sum of their salvage values. The group's useful life is the average of the lives of all the items. Any of the unit depreciation methods can be used to compute the group depreciation.

Example 11.5 The ABC Company purchased five cutters with useful lives of 5, 6, 7, 8, and 10 years; the cutters cost $10 000, $12 000, $13 000, $14 000, and $16 000, respectively. The salvage value of each cutter is estimated to be $500. Compute the annual group depreciation charge, using the straight-line method.

$$\text{group adjusted cost} = (\$10\,000 + \cdots + \$16\,000) - 5(\$500) = \$62\,500$$

$$\text{group useful life} = \frac{5+6+7+8+10}{5} = 7.2 \text{ years}$$

$$\text{group SD} = \frac{\$62\,500}{7.2} = \$8680.56$$

When a mixed collection of assets is subdivided into groups according to useful life, one can perform a *composite depreciation.* First, an annual depreciation charge is calculated for each life group by use of a group depreciation method. The sum of these charges then gives the composite annual depreciation. A composite life, n, can now be *defined* as the total of all adjusted costs, divided by the composite depreciation amount. The corresponding composite depreciation rate is defined as $100\%/n$.

Example 11.6 Rework Example 11.5 by composite depreciation.

Applying the straight-line method to each life group, which in this case consists of a single item, we obtain:

$$\text{SD}_{\text{5-group}} = \frac{\$10\,000 - \$500}{5} = \$1900.00$$

$$\text{SD}_{\text{6-group}} = \frac{\$12\,000 - \$500}{6} = \$1916.67$$

$$\text{SD}_{\text{7-group}} = \frac{\$13\,000 - \$500}{7} = \$1785.71$$

$$\text{SD}_{\text{8-group}} = \frac{\$14\,000 - \$500}{8} = \$1687.50$$

$$\text{SD}_{\text{10-group}} = \frac{\$16\,000 - \$500}{10} = \$1550.00$$

Then,

$$\text{composite SD} = \$1900.00 + \cdots + \$1550.00 = \$8839.88$$

$$\text{composite life} = \frac{62\,500}{8839.88} = 7.07 \text{ years}$$

$$\text{composite depreciation rate} = \frac{100\%}{7.07} = 14.14\%$$

Comparing with Example 11.5, we see that, under composite depreciation, a little more is charged off each year for a slightly smaller number of years.

The composite depreciation rate (or useful life) is restricted by guidelines issued by the U.S. Treasury Department. Variations from these guidelines must be individually approved by the Internal Revenue Service.

11.7 ADDITIONAL FIRST-YEAR DEPRECIATION; INVESTMENT TAX CREDIT

The U.S. government from time to time seeks to encourage new capital investments by business, to stimulate the economy. Two incentives have been used, either of which may be enacted into law when the federal government feels stimuli are needed, and withdrawn when it feels stimuli are not needed.

The *additional first-year depreciation* provision allows an additional percentage depreciation deduction during the year in which the asset was purchased. The percentage is based on the original cost of the asset, and is in addition to any regular depreciation. The *investment tax credit* provision allows a business to reduce its annual income tax by some stated percentage of the original cost of any assets purchased during that year.

Example 11.7 The SSG Company spent $1 000 000 for new equipment on January 1 of this year. The equipment has a useful life of 10 years, zero salvage value, and is depreciated by the straight-line method. Additional first-year depreciation of 20% and an investment tax credit of 7% apply. Compute the total first-year depreciation, depreciation for other years, and the first-year investment tax credit for the SSG Company.

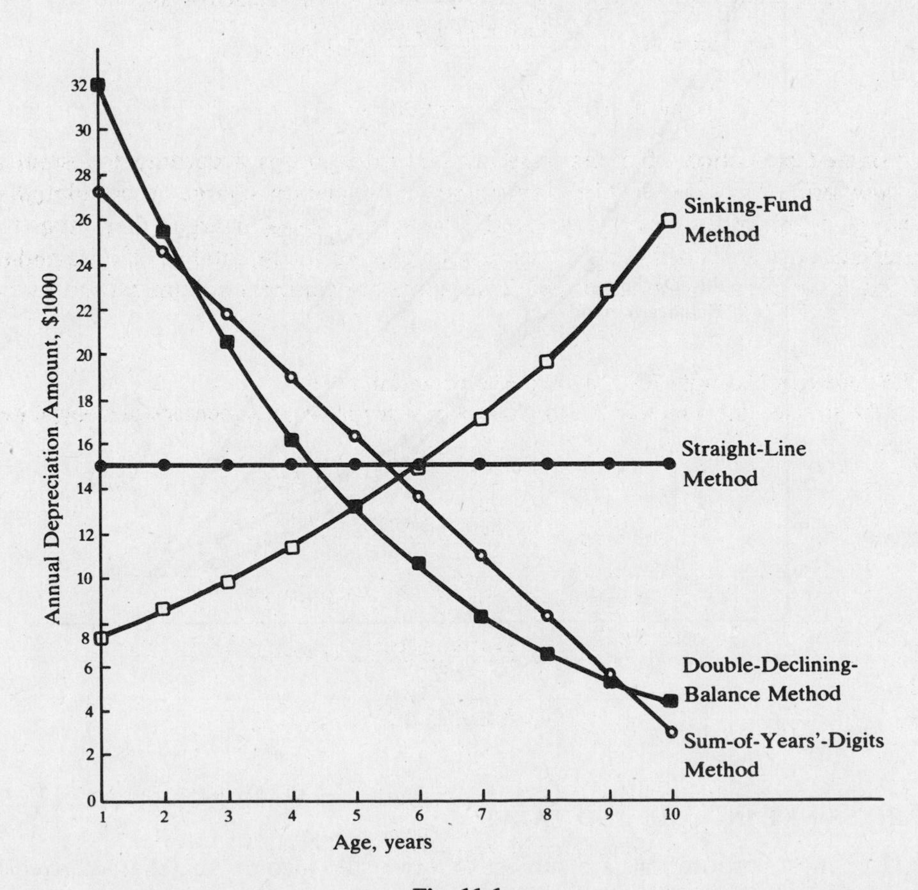

Fig. 11-1

We have
$$SD = \frac{\$1\,000\,000}{10} = \$100\,000$$

This year, the depreciation charge will be
$$S_1 = \$100\,000 + (0.20)(\$1\,000\,000) = \$300\,000$$

Since 3 SD is charged off in year 1, the annual depreciation will be SD in years 2 through 8, and zero in years 9 and 10. In addition, the SSG Company can reduce this year's tax bill by
$$(0.07)(\$1\,000\,000) = \$70\,000$$

11.8 COMPARISON OF DEPRECIATION METHODS

Figures 11-1 and 11-2 are plots of the annual depreciation charges and book values from Tables 11-1, 11-2, 11-4, and 11-5. Figure 11-1 makes manifest what we have said about the four traditional depreciation schemes: the sum-of-years'-digits method and the declining-balance methods are *accelerated* (heaviest depreciation in earlier years); the sinking-fund method is *decelerated* (heaviest depreciation in later years); and the straight-line method is neither accelerated nor decelerated. Notice, in Fig. 11-2, that the book values generated by the straight-line method are intermediate between those for the accelerated and the decelerated methods. This holds true in general.

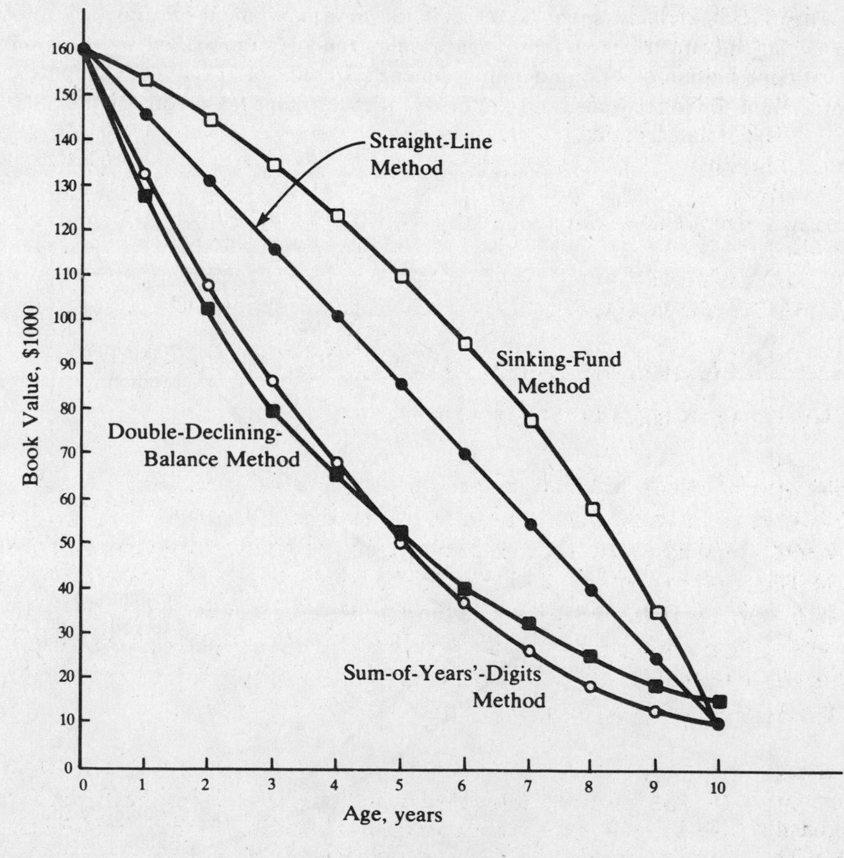

Fig. 11-2

11.9 BUSINESS NET INCOME AND TAXES

In the U.S., most corporations are subject to a two-step income tax, characterized by a *base tax rate* and a *surtax rate*. The base tax rate, r_t, is currently 22% of net taxable income, I, where

$$I = (\text{gross receipts and sales}) - (\text{bad debts}) - (\text{cost of goods sold}) - (\text{wages} + \text{salaries}) - (\text{interest} + \text{rent}) - \text{depreciation} \tag{11.15}$$

The surtax rate, s_t, is presently 26%; it applies only to income in excess of \$25 000. Total corporate income tax is thus given by

$$T = Ir_t + (I - \$25\,000)s_t \tag{11.16}$$

where it is assumed that $I \geq \$25\,000$. For most corporations, $I \gg \$25\,000$, so that

$$T \approx t_r I \tag{11.17}$$

where $t_r = r_t + s_t = 48\%$.

Unincorporated businesses are generally taxed at the individual tax rate(s) of the owner(s).

Example 11.8 The KJL Corporation received \$10 000 000 from the sales of their products during the current year. A total of \$1000 of these sales was never actually collected and was accounted for as bad debts. The company spent \$3 000 000 in the production and warehousing of their products during the current year. A total of \$1 000 000 was spent for wages and salaries, \$500 000 was paid out in interest on long-term loans, \$700 000 was spent for rental of space and equipment, and \$600 000 depreciation was charged off. Compute the KJL Corporation's income tax bill for the current year, if the base tax rate is 22% and the surtax rate is 26%.

From (*11.15*):

Gross Receipts and Sales	=	\$10 000 000	
Less: Bad Debts	=	1 000	
Gross Income	=	\$ 9 999 000	
Period Costs:			
Cost of Goods Sold	=		\$3 000 000
Wages and Salaries	=		1 000 000
Interest	=		500 000
Rent	=		700 000
Less: Total Period Costs	=	5 200 000	
Net Income	=	\$ 4 799 000	
Less: Depreciation	=	600 000	
Net Taxable Income	=	\$ 4 199 000	

$$T = (\$4\,199\,000)(0.22) + (\$4\,199\,000 - \$25\,000)(0.26) = \$2\,009\,020$$

[or, by (*11.17*), $T \approx 0.48 \times \$4\,199\,000 = \$2\,015\,520$].

Capital gains (*losses*) occur when an asset is sold for more (less) than its book value. Under current U.S. tax laws, if an asset has been held more than six months, it is a *long-term* asset; otherwise, it is a *short-term* asset. Long-term capital gains and losses are aggregated separately from short-term capital gains and losses. If the long-term aggregate is positive (a capital gain), it is taxed at only 30% (not at $t_r = 48\%$). If the long-term aggregate is negative (a capital loss), this loss may be carried forward and spread arbitrarily over the next five years, as an offset to any capital gains during those years. If the short-term aggregate is positive (a capital gain), it is taxed as regular income, at t_r; if it is negative (a capital loss), it is treated like a long-term capital loss.

Example 11.9 Assume that in the current year the KJL Corporation sells a machine, which it has used for several years, for \$600 000. The machine originally cost \$500 000 and has been depreciated under the double-declining-balance method; the current book value is \$300 000. The corporation also had short-term capital losses of \$50 000 and short-term capital gains of \$20 000. Compute the KJL Corporation's income tax bill, using the other information in Example 11.8.

Sale of the machine results in a *long-term capital gain* of

$$\$600\,000 - \$500\,000 = \$100\,000$$

plus *ordinary income* in the amount

$$(\$600\,000 - \$300\,000) - \$100\,000 = \$200\,000$$

In addition, there is an aggregate short-term capital loss of \$30 000. KJL will therefore have to pay taxes of

$$\$2\,009\,020 + (0.48)(\$200\,000) + (0.30)(\$100\,000) = \$2\,135\,020$$

but may claim a credit of \$30 000 against future capital gains.

11.10 COMPARATIVE EFFECTS OF DEPRECIATION METHODS ON INCOME TAXES

Depreciation is deducted as an expense of doing business. Thus, by lowering net income, depreciation lowers income taxes. Specifically, if the normal tax rate is t_r, then the *depreciation tax shield* in year j, or amount of taxes saved in that year because depreciation is taken, is

$$S_j = t_r D_j \tag{11.18}$$

where D_j is the depreciation charge for that year.

Over the life of an asset, the total amount of depreciation tax shield will be the same under the straight-line, sum-of-years'-digits, and sinking-fund methods (since each of these methods charges off the entire adjusted cost of the asset). The declining-balance method will often yield a slightly smaller depreciation tax shield over the life of the asset, which, however, will be more or less compensated for by the capital loss suffered when the asset is sold at a salvage value below its book value. Thus, all four methods will give approximately the same *total* tax shield over the useful life of the asset. However, because of the time value of money, accelerated depreciation methods (declining-balance and sum-of-years'-digits) will yield a larger present-worth net income after taxes. This is because the accelerated methods provide a larger tax shield in the earlier years of the asset's life; and the earlier the savings, the less they are discounted in calculating the present worth.

Example 11.10 Assume that the tax rate is 52% and the (before-tax) net income is \$100 000 per year, before depreciation. Compare the effects of the four depreciation methods from (*a*) Example 11.1, (*b*) Example 11.2(*a*), (*c*) Example 11.3, (*d*) Example 11.4. Assume MARR = 15%.

(*a*)

annual taxes = (0.52)(\$100 000 − \$15 000) = \$44 200
total taxes for 10 years = \$442 000
present worth of 10 years' taxes = \$44 200(P/A, 15%, 10) = \$221 831.87

(*b*) The tax shield from the capital loss (at maximum rate) is \$7179.87 (0.30) = \$2153.96.

Table 11-6

Year	Net Taxable Income, I	Taxes, $0.52I$	Present Worth of Taxes
1	\$100 000 − \$32 000 = \$68 000.00	\$35 360.00	\$30 747.83
2	100 000 − 25 600 = 74 400.00	38 688.00	29 253.69
3	100 000 − 20 480 = 79 520.00	41 350.40	27 188.56
4	100 000 − 16 384 = 83 616.00	43 480.32	24 860.01
5	100 000 − 13 107.20 = 86 892.80	45 184.26	22 464.56
6	100 000 − 10 485.76 = 89 514.24	46 547.41	20 123.73
7	100 000 − 8388.61 = 91 611.39	47 637.92	17 908.86
8	100 000 − 6710.88 = 93 289.12	48 510.34	15 858.12
9	100 000 − 5368.71 = 94 631.29	49 208.27	13 988.06
10	100 000 − 4294.96 = 95 705.04	49 766.62	12 301.55
	TOTALS	\$445 735.54	\$214 694.97

Then, from Table 11-6,

$$\text{adjusted total taxes for 10 years} = \$445\,735.54 - \$2153.96 = \$443\,581.58$$

$$\text{present worth of 10 years' taxes} = \$214\,694.97$$

$$\text{adjusted present worth of 10 years' taxes} = \$214\,694.97 - \$2153.96(P/F, 15\%, 10) = \$214\,162.54$$

(*c*) The 10 years' taxes and their present worth are given by the totals in Table 11-7.

Table 11-7

Year	Net Taxable Income, I	Taxes, $0.52I$	Present Worth of Taxes
1	\$100 000 − \$27 272.73 = \$72 727.27	\$ 37 818.18	\$ 32 885.37
2	100 000 − 24 545.45 = 75 454.55	39 236.37	29 668.33
3	100 000 − 21 818.18 = 78 181.82	40 654.55	26 731.02
4	100 000 − 19 090.91 = 80 909.09	42 072.73	24 055.31
5	100 000 − 16 363.64 = 83 636.36	43 490.91	21 622.67
6	100 000 − 13 636.36 = 86 363.64	44 909.09	19 415.44
7	100 000 − 10 909.09 = 89 090.91	46 327.27	17 416.14
8	100 000 − 8181.82 = 91 818.18	47 745.45	15 608.07
9	100 000 − 5454.55 = 94 545.45	49 163.63	13 975.37
10	100 000 − 2727.27 = 97 272.73	50 581.82	12 503.05
	TOTALS	\$442 000.00	\$213 880.77

(*d*) The 10 years' taxes and their present worth are given by the totals in Table 11-8.

Table 11-8

Year	Net Taxable Income, I	Taxes, $0.52I$	Present Worth of Taxes
1	\$100 000 − \$ 7387.81 = \$92 612.19	\$ 48 158.34	\$ 41 876.82
2	100 000 − 8495.98 = 91 504.02	47 582.09	35 978.90
3	100 000 − 9770.38 = 90 229.62	46 919.40	30 850.27
4	100 000 − 11 235.93 = 88 764.06	46 157.31	26 390.59
5	100 000 − 12 921.32 = 87 078.67	45 280.91	22 512.62
6	100 000 − 14 859.52 = 85 140.48	44 273.05	19 140.46
7	100 000 − 17 088.45 = 82 911.55	43 114.00	16 208.15
8	100 000 − 19 651.72 = 80 348.28	41 781.10	13 658.32
9	100 000 − 22 599.48 = 77 400.52	40 248.27	11 441.07
10	100 000 − 25 989.40 = 74 010.60	38 485.51	9 513.03
	TOTALS	\$441 999.98	\$227 570.23

Comparing the above results, we see that the *total* taxes for 10 years are \$442 000 under each depreciation method (with a deviation of \$1581.58 for the double-declining-balance method). However, on a present-worth basis we have:

Method	Present-Worth Tax Advantage (+) or Disadvantage (−) Relative to Straight-Line Method
Double-Declining-Balance	+\$7669.33
Sum-of-Years'-Digits	+\$7951.18
Sinking-Fund	−\$5738.36

11.11 THE ACCELERATED COST RECOVERY SYSTEM

The Accelerated Cost Recovery System (ACRS) is a depreciation method recently instituted by the Internal Revenue Service. It is mandatory for most tangible assets placed in service after December 31, 1980. Its main features are that salvage value is not relevant and that the useful life of the asset is limited to 3, 5, 10, or 15 years. The IRS publishes depreciation scales for each class life. Thus, the percentages for *three-year property* placed in service during 1982 are 25% for the first year, 38% for the second year, and 37% for the third year. This class life includes assets with a useful life of 4 years or less, such as automobiles, small trucks, and some manufacturing tools. Items used in research and experimentation are also included in this category.

Five-year property includes office furniture, some storage facilities, and, in general, all property that is not three-, ten-, or fifteen-year. The percentages are 15% for the first year; 22% for the second year; and 21% for the third, fourth, and fifth year.

Ten-year property includes assets with a useful life of less than 12.5 years. The percentages are 8% for the first year, 14% for the second year, 12% for the third year, 10% for each of years four through six, and 9% for each of years seven through ten.

Assets with a useful life of more than 12.5 years are designated as *fifteen-year property.* There is one rate structure for low-income housing and another for all other fifteen-year property. Percentages for the fifteen-year asset also depend on the month the property was placed in service; e.g., for an asset (not low-income housing) placed in service in April, the percentages are:

Year	1	2	3	4	5	6–10	11–15	16
%	9	11	9	8	7	6	5	1

Example 11.11 Apply the ACRS to the asset of Example 11.1.

The calculations for this ten-year asset are given in Table 11-9.

Table 11-9

Year	Depreciation Rate for Year	Depreciation Charge for Year	Accumulated Depreciation	Book Value at End of Year
1	8%	\$12 800	\$ 12 800	\$147 200
2	14	22 400	35 200	124 800
3	12	19 200	54 400	105 600
4	10	16 000	70 400	89 600
5	10	16 000	86 400	73 600
6	10	16 000	102 400	57 600
7	9	14 400	116 800	43 200
8	9	14 400	131 200	28 800
9	9	14 400	145 600	14 400
10	9	14 400	160 000	0

Example 11.12 Assume again, as in Example 11.10, and tax rate of 52% and an annual income of $100 000 before depreciation and taxes. Compare the effect of the ACRS (using Table 11-9) with those of the four traditional methods, as found in Example 11-10.

Table 11-10

Year	Net Taxable Income, I	Taxes, $0.52I$	Present Worth of Taxes (i = 15%)
1	$100 000 − $12 800 = $87 200	$45 344	$39 429.57
2	100 000 − 22 400 = 77 600	40 352	30 511.91
3	100 000 − 19 200 = 80 800	42 016	27 626.20
4	100 000 − 16 000 = 84 000	43 680	24 974.18
5	100 000 − 16 000 = 84 000	43 680	21 716.68
6	100 000 − 16 000 = 84 000	43 680	18 884.07
7	100 000 − 14 400 = 85 600	44 512	16 733.71
8	100 000 − 14 400 = 85 600	44 512	14 551.05
9	100 000 − 14 400 = 85 600	44 512	12 653.09
10	100 000 − 14 400 = 85 600	44 512	11 002.09
	TOTALS	$436 800	$218 082.55

See Table 11-10. The comparison made in Example 11.10 may now be extended as follows:

Method	*Present-Worth Tax Advantage (+) or Disadvantage (−) Relative to Straight-Line Method*
Double-Declining-Balance	+$7669.33
Sum-of-Years'-Digits	+$7951.18
Sinking-Fund	−$5738.36
ACRS	+$3748.72

It is seen that, in this case, the ACRS is better than straight-line depreciation but not as good as the two traditional accelerated methods. However, had the useful life of the asset been 12 years, the ACRS would have been best (the asset would still be ten-year property, but the traditional methods would have to be applied over the whole 12 years).

11.12 CHOICE OF DEPRECIATION METHOD

As indicated, a taxpayer would have to present excellent arguments (based on facts, not opinions) to be allowed to use a method other than ACRS for tangible assets. For intangible assets (franchises, designs, drawings, copyrights, patterns, subscription lists, customer lists, etc.), the traditional methods can still be applied, singly or in combination, provided the useful life can be projected with reasonable accuracy.

As can be seen from Fig. 11-2, the use of the double-declining-balance during the early years, combined with a switch to the sum-of-years'-digits in later years, may provide the greatest possible tax shield, for this combination gives the largest possible accumulated depreciation charge (smallest possible book value) over the life of the asset. However, this combination may result in an accumulated depreciation at the end of the useful life of the asset which exceeds the asset's adjusted cost—a violation of tax regulations. Moreover, the current tax laws specifically prohibit certain switches in depreciation method without the prior approval of the Internal Revenue Service.

11.13 DEPRECIATION AND CASH FLOW

Depreciation, as an accounting charge against income, is not itself a cash flow. However, it does influence the amount of income tax paid, which is a (negative) cash flow. For year j, let us write:

$$\mathrm{BTCF}_j \equiv \text{before-tax net cash flow}$$
$$\mathrm{ATCF}_j \equiv \text{after-tax net cash flow}$$
$$I_j \equiv \text{net taxable income}$$
$$\mathrm{ATI}_j \equiv \text{after-tax net income}$$
$$T_j \equiv \text{income tax}$$
$$D_j \equiv \text{depreciation charge}$$

Then, $\mathrm{ATCF}_j = \mathrm{BTCF}_j - T_j$. But, by (*11.15*),

$$\mathrm{ATI}_j = I_j - T_j = (\mathrm{BTCF}_j - D_j) - T_j$$

Consequently,

$$\mathrm{ATCF}_j = \mathrm{ATI}_j + D_j \tag{11.19}$$

that is to say, the after-tax cash flow for the year is the sum of the after-tax net income for the year and the depreciation charge for the year.

11.14 BEFORE- AND AFTER-TAX ECONOMIC ANALYSES

Economic analyses should generally be made on an after-tax basis, unless it is clear that tax considerations are irrelevant. We have seen that depreciation can cause the before-tax and after-tax pictures to differ. Deduction of interest paid on borrowed money will have a similar effect. Perhaps most significant is the fact that businesses are judged (by analysts and investors) on the basis of their *after-tax* performance.

Example 11.13 The ABC Company is planning to buy a new pump. The pump costs \$50 000, and has a 10-year life and zero salvage value. The pump will increase the company's net income before taxes by \$12 000 in each of the 10 years. The company's tax rate is 51%. What is the ROR on the pump?

We make three different analyses, which give three different results.

Before-Tax

$$0 = -\$50\,000 + \$12\,000(P/A, i^*\%, 10) \qquad \text{whence} \qquad i^* = 20.2\%$$

After-Tax, No Depreciation

With $D_j = 0$, (*11.19*) gives $\mathrm{ATCF}_j = (0.49)(\$12\,000) = \$5880$; thus,

$$0 = -\$50\,000 + \$5880(P/A, i^*\%, 10) \qquad \text{whence} \qquad i^* = 3\%$$

After-Tax, Straight-Line Depreciation

With $D_j = \$50\,000/10 = \$5000$, (*11.19*) gives

$$\mathrm{ATCF}_j = (0.49)(\$12\,000 - \$5000) + \$5000 = \$8430$$

and so

$$0 = -\$50\,000 + \$8430(P/A, i^*\%, 10) \qquad \text{whence} \qquad i^* = 10.9\%$$

Example 11.14 The ABC Company (Example 11.13) can purchase an alternative "Superpump" that costs \$100 000 and generates an annual increase in the company's net income before taxes of \$23 852. This pump also has a 10-year life and zero salvage value; however, a special provision in the tax laws permits depreciation of this pump over a 5-year period. Which pump should the company buy?

For the Superpump, we make two analyses:

Before-Tax

$$0 = -\$100\,000 + \$23\,852(P/A, i^*, 10) \qquad \text{whence} \qquad i^* = 20\%$$

After-Tax, Straight-Line Depreciation

For $j = 1, 2, \ldots, 5$:

$$D_j = \frac{\$100\,000}{5} = \$20\,000$$

$$\text{ATCF}_j = (0.49)(\$23\,852 - \$20\,000) + \$20\,000 = \$21\,887.48$$

while for $j = 6, 7, \ldots, 10$:

$$D_j = 0$$

$$\text{ATCF}_j = (0.49)(\$23\,852) = \$11\,687.48$$

$$0 = -\$100\,000 + \$21\,887.48(P/A, i^*\%, 5) + \$11\,687.48(P/A, i^*\%, 5)\,(P/F, i^*, 5)$$

which yields $i^* = 12.8\%$.

Comparing the above results with those of Example 11.3, we conclude that the Superpump is not to be preferred on a before-tax basis, but is definitely to be preferred on an after-tax basis. In such cases, the after-tax picture is always the correct one.

Solved Problems

11.1 A company's tax rate is 52%. To improve labor relations, the company has decided to donate $1 000 000 to its labor union to build a sports arena for the use of union members and the general public. (*a*) If the gift is ruled tax deductible, what is the actual cost to the company? (*b*) If the gift is ruled nondeductible, what is the actual cost to the company, and how does it account for the gift? (*c*) If the labor union is a tax-exempt corporation, to what extent is the general public (through the government) subsidizing the arena?

(*a*)

$$\text{tax savings} = (0.52)(\$1\,000\,000) = \$520\,000$$

$$\text{actual cost to the company} = \$480\,000$$

(*b*) $1 000 000; nondeductible expense.

(*c*) If the donation is ruled tax deductible, the general public would be footing the company's tax savings, $520 000. (If the labor union were not tax-exempt, it would have to pay taxes on the gift, in which case the public's subsidy would amount to $520 000 minus the union's taxes.)

11.2 A computer system can be purchased for $18 000. The operating costs will be $10 000 per year, and the useful life is expected to be 5 years, with $5000 salvage value at that time. The present annual sales volume should increase by $16 000 as a result of acquiring the computer system. The company's tax rate is 50%. (*a*) Depreciate the asset by the straight-line method. (*b*) Compute annual taxes, annual cash flows after taxes, and after-tax ROR for the investment.

(*a*) SD = ($18 000 − $5000)/5 = $2600; see Table 11-11.

Table 11-11

Year	Depreciation Charge for Year	Accumulated Depreciation	Book Value at End of Year
1	$2600	$ 2 600	$15 400
2	2600	5 200	12 800
3	2600	7 800	10 200
4	2600	10 400	7 600
5	2600	13 000	5 000

(*b*) The annual net cash flow before taxes is $16 000 − $10 000 = $6000; see Table 11-12.

Table 11-12

	Year 1	Year 2	Year 3	Year 4	Year 5
BTCF	$6000	$6000	$6000	$6000	$6000
Depreciation	2600	2600	2600	2600	2600
Net Taxable Income	3400	3400	3400	3400	3400
Tax (@ 50%)	1700	1700	1700	1700	1700
ATCF	4300	4300	4300	4300	4300

The after-tax ROR may be found by equating the after-tax PW to zero (see Chapter 7):

$$0 = -\$18\,000 + \$4300(P/A, i^*\%, 5) + \$5000(P/F, i^*\%, 5)$$

or $i^* = 12.6\%$.

11.3 Rework Problem 11.2 if the IRS rules that the equipment's *tax life* is eight years, with $2000 salvage value at that date.

(*a*) SD = ($18 000 − $2000)/8 = $2000; see Table 11-13.

Table 11-13

Year	Depreciation Charge for Year	Accumulated Depreciation	Book Value at End of Year
1	$2000	$ 2 000	$16 000
2	2000	4 000	14 000
3	2000	6 000	12 000
4	2000	8 000	10 000
5	2000	10 000	8 000

The useful life, or depreciation period, remains 5 years. At that time the computer system would presumably be sold—for $3000 less than its book value. Thus the company would take a long-term capital loss of $3000 and carry it forward to offset long-term capital gains over the next five years.

(*b*) See Table 11-14.

Table 11-14

	Year 1	Year 2	Year 3	Year 4	Year 5
BTCF	$6000	$6000	$6000	$6000	$6000
Depreciation	2000	2000	2000	2000	2000
Net Taxable Income	4000	4000	4000	4000	4000
Tax (@ 50%)	2000	2000	2000	2000	2000
ATCF	4000	4000	4000	4000	4000

Assuming that the company takes its tax credit [see (*a*)] in year 6, it will save (0.30)($3000) = $900 in long-term capital gains taxes in that year. Hence, the after-tax ROR is given by

$$0 = -\$18\,000 + \$4000(P/A, i^*\%, 5) + \$5000(P/F, i^*\%, 5) + \$900(P/F, i^*\%, 6)$$

or $i^* = 11.4\%$.

11.4 Rework Problem 11.2 using the double-declining-balance depreciation method.

(*a*) With $r_d = 2r_s = 200\%/5 = 40\%$, we generate Table 11-15.

Table 11-15

Year	Depreciation Charge for Year	Accumulated Depreciation	Book Value at End of Year
1	$7200	$ 7 200	$10 800
2	4320	11 520	6 480
3	1480	13 000	5 000
4	0	13 000	5 000
5	0	13 000	5 000

Observe that the third-year values had to be adjusted so that the accumulated depreciation would not exceed the maximum set by the IRS (the adjusted cost). No depreciation can be taken in years 4 and 5, and the book value of $5000 remains undepreciated.

(*b*) See Table 11-16.

Table 11-16

	Year 1	Year 2	Year 3	Year 4	Year 5
BTCF	$6000	$6000	$6000	$6000	$6000
Depreciation	7200	4320	1480	0	0
Net Taxable Income	−1200	1680	4520	6000	6000
Tax (@50%)	−600	840	2260	3000	3000
ATCF	6600	5160	3740	3000	3000

For the after-tax ROR:

$$0 = -\$18\,000 + \$6600(P/F, i^*\%, 1) + \$5160(P/F, i^*\%, 2)$$
$$+ [\$3740 + \$3000(P/A, i^*\%, 2)]\,(P/F, i^*\%, 3) + \$5000(P/F, i^*\%, 5)$$

or $i^* = 13.6\%$.

11.5 Rework Problem 11.2 using the sum-of-years'-digits method of depreciation.

(*a*) SY = (5)(6)/2 = 15; see Table 11-17.

Table 11-17

Year	Depreciation Charge for Year	Accumulated Depreciation	Book Value at End of Year
1	$(\frac{5}{15})(\$13\,000) = \$4333.33$	$ 4 333.33	$13 666.67
2	$(\frac{4}{15})(13\,000) = 3466.67$	7 800.00	10 200.00
3	$(\frac{3}{15})(13\,000) = 2600.00$	10 400.00	7 600.00
4	$(\frac{2}{15})(13\,000) = 1733.33$	12 133.33	5 866.67
5	$(\frac{1}{15})(13\,000) = 866.67$	13 000.00	5 000.00

Table 11-18

	Year 1	Year 2	Year 3	Year 4	Year 5
BTCF	$6000.00	$6000.00	$6000	$6000.00	$6000.00
Depreciation	4333.33	3466.67	2600	1733.33	866.67
Net Taxable Income	1666.67	2533.33	3400	4266.67	5133.33
Tax (@50%)	833.33	1266.67	1700	2133.33	2566.67
ATCF	5166.67	4733.33	4300	3866.67	3433.33

(*b*) See Table 11-18.

In writing the equation for the after-tax ROR, we note that the depreciation charges form a gradient series, with

$$G = -\frac{1}{15}(\$13\,000) = -\$866.67$$

Hence, the ATCF_j also form a gradient series, with

$$G' = (0.50)(-\$866.67) = -\$433.33$$

and we have:

$$0 = -\$18\,000 + [\$5166.67 - \$433.33(A/G, i^*\%, 5)]\,(P/A, i^*\%, 5) + \$5000(P/F, i^*\%, 5)$$

or $i^* = 14.5\%$.

11.6 Rework Problem 11.2(*a*) using the sinking-fund method of depreciation and a before-tax MARR of 12%.

From (*11.13*), with C = adjusted cost = $13 000,

$$\text{SFD}_j = \$13\,000(A/F, 12\%, 5)(1.12)^{j-1} = (\$2046.33)(1.12)^{j-1}$$

and we obtain Table 11-19.

Table 11-19

Year	Depreciation Charge for Year	Accumulated Depreciation	Book Value at End of Year
1	$2046.33	$ 2 046.33	$15 953.67
2	2291.89	4 338.21	13 661.79
3	2566.91	6 905.12	11 094.88
4	2874.94	9 780.07	8 219.93
5	3219.93	13 000.00	5 000.00

11.7 Refer to Problems 11.2 and 11.6. Is the computer system a viable proposition if the company's after-tax MARR is 12%?

Table 11-20

	Year 1	Year 2	Year 3	Year 4	Year 5
BTCF	$6000.00	$6000.00	$6000.00	$6000.00	$6000.00
Depreciation	2046.33	2291.89	2566.91	2874.94	3219.93
Net Taxable Income	3953.67	3708.11	3433.09	3125.06	2780.07
Tax (@ 50%)	1976.84	1854.06	1716.55	1562.53	1390.04
ATCF	4023.16	4145.95	4283.45	4437.47	4609.96

The after-tax cash flows are computed in Table 11-20. Then, at MARR = 12%,

$$\begin{aligned} PW &= -\$18\,000 + \$4023.16(P/F, 12\%, 1) + \$4145.95(P/F, 12\%, 2) + \$4283.45(P/F, 12\%, 3) \\ &\quad + \$4437.47(P/F, 12\%, 4) + \$4609.96(P/F, 12\%, 5) + \$5000(P/F, 12\%, 5) \\ &= \$219.18 \end{aligned}$$

From this, we conclude that the investment meets the after-tax MARR and pays an extra $219.18 (in today's money) over the five-year useful life.

11.8 Rework Problem 11.2 using the ACRS. Assume that the IRS classifies the asset as five-year property.

(*a*) See Table 11-21.

Table 11-21

Year	Depreciation Rate for Year	Depreciation Charge for Year	Accumulated Depreciation	Book Value at End of Year
1	15%	$2700	$ 2 700	$15 300
2	22	3960	6 660	11 340
3	21	3780	10 440	7 560
4	21	3780	14 220	3 780
5	21	3780	18 000	0

(*b*) See Table 11-22.

Table 11-22

	Year 1	Year 2	Year 3	Year 4	Year 5
BTCF	$6000	$6000	$6000	$6000	$6000
Depreciation	2700	3960	3780	3780	3780
Net Taxable Income	3300	2040	2220	2220	2220
Tax (@ 50%)	1650	1020	1110	1110	1110
ATCF	4350	4980	4890	4890	4890

Sale of the equipment for its salvage value at the end of year 5 produces a long-term capital gain, of which the value after taxes is

$$(1-0.30)(\$5000)=\$3500$$

Hence, the equation for the after-tax ROR is

$$0=-\$18\,000+\$4350(P/F,i^*\%,1)+\$4980(P/F,i^*\%,2)$$
$$+\$4890(P/A,i^*\%,3)(P/F,i^*\%,2)+\$3500(P/F,i^*\%,5)$$

giving $i^*=14.45\%$.

11.9 Rework Problem 11.8 on the assumption that the asset is reclassified as three-year property.

(*a*) See Table 11-23.

Table 11-23

Year	Depreciation Rate for Year	Depreciation Charge for Year	Accumulated Depreciation	Book Value at End of Year
1	25%	$4500	$ 4 500	$13 500
2	38	6840	11 340	6 660
3	37	6660	18 000	0
4	0	0	18 000	0
5	0	0	18 000	0

(*b*) See Table 11-24.

Table 11-24

	Year 1	Year 2	Year 3	Year 4	Year 5
BTCF	$6000	$6000	$6000	$6000	$6000
Depreciation	4500	6840	6660	0	0
Net Taxable Income	1500	−840	−660	6000	6000
Tax (@50%)	750	−420	−330	3000	3000
ATCF	5250	6420	6330	3000	3000

$$0=-\$18\,000+\$5250(P/F,i^*\%,1)+\$6420(P/F,i^*\%,2)+\$6330(P/F,i^*\%,3)$$
$$+\$3000(P/A,i^*\%,2)(P/F,i^*\%,3)+\$3500(P/F,i^*\%,5)$$

whence $i^*=16.27\%$. Note the benefit to the taxpayer when the asset's cost is depreciated over a period shorter than the useful life.

Supplementary Problems

11.10 For the asset of Problem 11.2, (*a*) determine the ROR of the project before taxes and (*b*) recommend a depreciation method on the basis of Problems 11.2(*b*), 11.4(*b*), 11.5(*b*), 11.7, and 11.8(*b*).
Ans. (*a*) 24.4%; (*b*) sum-of-years'-digits (if allowed by the IRS, which is doubtful)

11.11 A car rental agency has bought three economy-size, four medium-size, and two full-size cars; see Table 11-25. Using composite depreciation, compute the annual straight-line depreciation charge and the (composite) life for this collection of cars. *Ans.* \$12 100, 324/121 = 2.678 years

Table 11-25

	Economy	Medium	Full
Initial Cost (each)	\$6200	\$8200	\$13 000
Service Life	2 years	3 years	3 years
Salvage Value (each)	\$3600	\$4900	\$7300

11.12 Rework Problem 11.11 using group depreciation. *Ans.* \$12 150, 8/3 = 2.667 years

11.13 Rework Problem 10.16, assuming a 4-year tax life remaining for the current machine and a 4-year tax life for the new one. Straight-line depreciation is used, and an after-tax MARR of 15% is applicable. The tax rate is 50%. *Ans.* after-tax EUAC_{old} = \$6565, after-tax EUAC_{new} = \$7640; keep old machine.

11.14 Would the decision made in Problem 10.5 change if the second-hand machine could be depreciated in 10 years by the sum-of-years'-digits method and the company's tax rate is 52%? The after-tax MARR is 10%. *Ans.* after-tax $\text{EUAC}_{\text{shut-down}}$ = \$168, after-tax $\text{EUAC}_{\text{stand-by}}$ = \$407; decision unchanged.

11.15 Rework Problem 10.15, using an after-tax MARR of 10% and straight-line depreciation. The company's tax rate is 52%. *Ans.* 18 days

11.16 Would the economic life of the challenger in Problem 10.19 change if the sum-of-years'-digits depreciation method is used, the tax life is 8 years, the tax rate is 52%, and the after-tax MARR is 10%? *Ans.* no [after-tax EUAC(6) = \$2321]

11.17 Perform an after-tax analysis for the situation described in Problem 10.20. Assume that the service life and the tax life are identical, that straight-line depreciation is used, that the after-tax MARR is 10%, and that the company's tax rate is 50%. *Ans.* after-tax CUV > \$1000; keep current machine.

11.18 A machine's current book value is \$600. The machine cost \$1300 three years ago. Operating expenses have been \$380 per year, and the machine could last for three more years. Because of a breakthrough in design, a replacement machine which would save \$300 per year sells for \$1000 and has an expected service life of eight years. The scrap value of either machine at any time after installation is \$100. If the IRS allows straight-line depreciation over a six-year period for this type of machinery and if an after-tax MARR of 12% is acceptable, should the current machine be changed? The company's tax rate is 52%. *Ans.* after-tax $\text{EUAC}_{\text{current}}$ = \$90, after-tax $\text{EUAC}_{\text{replacement}}$ = \$167; keep current machine.

11.19 An income-producing asset costs \$60 000, has an estimated useful life of 7 years, has no salvage value after installation, and is expected to produce annual net savings of \$15 000. The company's tax rate is 52%. Compute (*a*) the before-tax ROR, and (*b*) the after-tax ROR under straight-line depreciation. *Ans.* (*a*) 16.3%; (*b*) 8.6%

11.20 For the situation described in Problem 11.19, compute the after-tax ROR when sum-of-years'-digits depreciation is charged. *Ans.* 9.7%

11.21 Fifty percent of the asset described in Problem 11.19 was financed from capital borrowed at 7%. This loan is to be repaid at the end of the seventh year, but interest is due on the principal at the end of each year. Rework Problem 11.20.
Ans. 26.0% (notice the big difference made by tax-deductible interest and the delay of seven years in half of the investment)

11.22 Would the result of Problem 10.23 change under straight-line depreciation, a 35% tax rate, and an after-tax MARR of 10%? Assume that the current machine is fully depreciated. *Ans.* no

11.23 Rework Problem 10.24 using straight-line depreciation, a 50% tax rate, and an after-tax MARR of 10%. Assume that salvage values are book values and that overhaul expenditures are depreciable.
Ans. Now the best action is to overhaul the old machine.

11.24 An underwater camera is purchased for $1000; it has an expected life of 12 years, at the end of which the estimated salvage value is $730. Using straight-line depreciation, find the book value of the camera at the end of 8 years. *Ans.* $420

11.25 An engineer is being transferred to another state and must vacate her house. The house, bought for $40 000 eight years ago, can be sold now for $60 000; the property is free of debt. If the house is sold, she will have to pay a 15% capital gains tax. The engineer is also considering leasing the house for five years, receiving $7200 annual rental. In this case, she estimates an annual disbursement of $1800 for taxes, insurance, and maintenance. She would also be allowed $1200 per year depreciation on her tax return (in addition to her cash disbursements); her rental income would be taxed at 30%. (Note that houses cannot be depreciated during the years used as the owner's personal residence.) Determine her after-tax ROR if she decides to lease the property and if the property shall be worth $64 000 at the end of the lease. *Ans.* 8%

11.26 A truck was bought 10 years ago for $70 000; its current salvage value is $14 000. It is believed that it can last 5 more years, at which time its salvage value will be $8000. Its operating expenses amount to $14 000 per year and they are expected to remain at that level for the next 5 years. The truck is currently being depreciated by the straight-line method, using a 15-year life and estimated salvage value of $10 000. A new truck can be purchased for $65 000. It will have yearly operating costs of $9000 and would last 20 years; salvage value after 20 years is estimated at $15 000. Again, straight-line depreciation would be used. Supposing that tax rates are 50% on income and 15% on capital gains (or losses), and that the company's after-tax MARR is 10%, should it buy the new truck?
Ans. No: after-tax $\text{EUAC}_{\text{old}} = \$7967$, after-tax $\text{EUAC}_{\text{new}} = \$10\,623$

11.27 A $60 000 asset will be depreciated by the straight-line method over a six-year period. No salvage value is expected. If the company's tax rate is 50%, what would be the present-worth advantage of using the sum-of-years'-digits method, given a 10% after-tax MARR? *Ans.* $1184

11.28 Rework Problem 11.26 using the ACRS for the new truck, classified as a 10-year asset. The current truck is still depreciated by the straight-line method.
Ans. No: after-tax $\text{EUAC}_{\text{old}} = \$7967$, after-tax $\text{EUAC}_{\text{new}} = \$9481$

11.29 Rework Problem 11.27 using the ACRS, with a five-year life, instead of the sum-of-years'-digits method.
Ans. −$3217 (i.e., straight-line depreciation is better in this case)

Chapter 12

Preparing and Presenting an Economic Feasibility Study

12.1 INTRODUCTION

This chapter attempts to bridge the gap between the specific analytic techniques—equivalent uniform annual series, rate of return, etc.—and the "wide-angle" considerations that determine (or should determine) investment decisions in the real world. Our frame of reference will be the presentation of a feasibility analysis to a lending institution.

Example 12.1 It will be assumed throughout this chapter that the project under study can be characterized as "marginal" with respect to the overall economic environment. Discuss the need for such an assumption and give examples of marginal projects.

A project is "marginal" if it will not significantly alter the economic environment. A dry-cleaner outlet in a major shopping center, a small die-casting factory in an industrial park, a new boutique in a fashion mall, are typical examples. Although undeniably important to their promoters, these ventures will not effect major changes in the economic patterns of the communities where they are to be implemented. For such ventures, clear-cut decisions may be derived from an engineering economic analysis.

By contrast, projects which represent a structural investment for the community (which would, for example, significantly alter the unemployment level or the gross regional product) would need in-depth study in every aspect and a thorough sensitivity analysis of each assumption underlying the feasibility study. Most likely, final decisions would be largely political in nature. Two such nonmarginal projects were Walt Disney World, which changed central Florida from a depressed rural area to a tourist capital, and the Trans-Alaska Oil Pipeline.

The sections that follow will discuss, one at a time, the main components of the feasibility study.

12.2 BACKGROUND INFORMATION

This section of the report should consist of:

1. A brief summary of the project, covering the nature of the venture, location, expected site, life, overall capital costs, financing, and return on investment
2. A description of the promoting individuals or institution: name(s), addresses (both legal and of proposed facility location), and institution characteristics (capital, number of shares of stock, principal shareholders, etc.)

Example 12.2 Why do most lending institutions require background information of the above type?

The lender will charge interest; nevertheless, he is risking his money, and before committing himself, understandably wants to gauge the risk as accurately as possible. Complete details about the board and executive officers of the promoting company, any partnerships or relationships to other companies, insurance available, and—most important—a history of past and current projects, are all commonly required. For industries currently in operation, data about their capacity, production level, productivity indicators, sales, labor force, salary structure, overhead, inventory turnover ratio, and other items (which may include some seemingly unrelated to the project itself) may be in order. Sometimes the lender requires an organizational chart, as well as details about production planning and control, quality control, labor relations, and financial situation (end of year balances for the last few years).

12.3 MARKET STUDY

In this section the report describes what is to be produced, and where and to whom it is going to be sold. An analysis should be made of:

1. The product—description, brand or name, quality standards to be met, characteristics, and utilization. Subproducts (if any)—description, utilization (if it is not going to be marketed).
2. The market—estimated demand for the final product. If the product can serve as raw material or intermediate product for some other article (e.g., a microprocessor to be used in an electronic toy), an estimate of potential uses and corresponding demands is in order. An analysis of complementary and competitive products, as well as their current and expected availabilities, should also be included.

Example 12.3 Discuss the importance of a thorough, realistic market study. What should be included in it, and what should be its ultimate objective? Show a typical summary output of this phase of the analysis.

A number of otherwise carefully planned ventures have failed because of an unjustified belief that the market will buy whatever can be produced. An excessive capacity is then built to exploit the good idea—so good, in fact, that it attracts immediate competition, and the long-term market share turns out to be lower than predicted. (Case in point: some fast-food retailers.)

A marketing study should provide information about: volume of similar products sold in the target region over the last, say, five years; local production; imports from, and exports to, other areas; consumption. It should give a quantification (and causes) of any unmet demand. An analysis of main producers is a must: location, market share, other areas served by them. Also, an analysis of main consumers: where they are located, historical consumption, uses of our product.

The goal of the market study is to provide the basis for a forecast of demand for the product in the first (say) five years after the operations start up. Not just a single number, but a range of values should be sought. A *pessimistic*, an *optimistic*, and a *most probable value* would be invaluable in a sensitivity analysis, provided the technical bases of the forecast are sound. A description of the techniques used in forecasting, as well as of the assumptions and data bases used in the analysis, should be included in an appendix to the report.

Figure 12-1 suggests the summary format.

	Year 1 (units)	Year 2 (units)	Year 3 (units)
1. Estimate of Regional Production			
2. Imports			
3. Exports			
4. Estimated Demand (1 + 2 − 3)			
With respect to the company			
5. Estimated Production			
6. Imports			
7. Exports			
8. Regional Demand Met (5 + 6 − 7)			

Fig. 12-1

3. Raw materials analysis. [Too often this phase of the study is replaced by an assumption that whatever is required will be there when (and in the amount) needed.]

Example 12.4 Discuss the potential dangers of omitting a raw materials analysis. Describe what such an analysis would yield. Give an example of obvious need for this type of study.

Unreliable providers may cause shortages, or they may force operators to keep excessive raw materials inventories lest they expose themselves to production stoppages. A raw materials analysis should include a study of prices (in the national and international markets), with considerations of transportation, insurance, tariffs, quality and delivery reliability of provider included for the latter and normal commercialization channels. If the project size is large, its potential effect on the raw materials market should be also considered. The availability of alternative providers, or even of alternative materials, may also turn out to be an important factor.

An extreme instance of the necessity for raw materials analysis was provided by the oil crisis of the 1970s.

12.4 PROJECT ENGINEERING

This section makes an analysis of available technologies for the process. A brief technical/economic comparison should be made and a justification of the selected technology provided. A statement of the potential consequences of this selection, a comparison with the "normal" competitor, as well as with the "state of the art" competitor, are desirable. Opinions from outside consultants, if available, regarding the selection of technology are also recommended.

Example 12.5 Describe the contents of the typical project engineering segment of the report. Comment on the effect of technology, size, start-up, and major equipment. Illustrate a typical summary output of this phase of the analysis.

The project engineering section of the economic feasibility report should include a thorough description of:

Fabrication process. No detail plans are needed; a flow diagram with durations of the stages, capacities, yields, and material and energy balances is normally sufficient for most manufacturing industries.

Project size. The planned production capacities must be forecast, indicating expected dates to attain them. Operating conditions (shifts per day, days per year) must be indicated. The relationship between this production plan and the market and raw materials studies should be included. An analysis of the market-share penetration must be provided. A justification of size from the point of view of the selected technology, financing limitations, plant location, seasonality factors, market restrictions, etc., is important. Analyses of size versus production costs, break-even point, resources needed to compensate for operating deficits over the start-up period, and impact of unforeseen slumps in sales are also required. A note should be made if future expansions to adapt to enlarged market share and product acceptance are envisioned.

Location. This extremely important factor is frequently lost in the shuffle. It is evident that remoteness of a location influences the pool of manpower available for the project, as well as the level of investment needed to provide housing, transportation, energy, water, sewage and tailings treatment, etc. These same factors should be considered for any location. Special consideration should be given to potential benefits related to the project's location, such as tax breaks, availability of cheap transportation, surrounding market, population, local regulations controlling noise and pollution, etc. The main factors influencing the selection of the project site should be discussed in detail, and an analysis of potential alternative sites, if any, should also be included.

Physical means of production. The plant site must be specified (total and build surfaces). Buildings required must also be detailed. Areas should be classified as direct production, ancillary facilities, administrative, warehouses, etc. If some of the buildings are already available and require little renovation, that should be noted. It is also very important to specify any earth moving, roads, docks, railroad connections, fences, etc., which should be needed as part of the site preparation. Note that electric substations, water treatment plant, and the like, are normally considered auxiliary facilities (see below).

Major equipment. Separate lists are usually made of domestic and imported equipment. The units are listed in order of decreasing value, until the total value of the two lists represents 60 to 70 percent of the entire capital investment. Numbers of the various units, their technical characteristics, theoretical capacities, prices (FOB; for imported equipment, extra charges such as transportation, in-transit insurance, tariffs, etc., must be detailed), and manufacturers are all indicated. Prices quoted should reflect actual market values, if at all possible. Pro forma invoices are of great help in documenting this phase of the analysis.

Auxiliary facilities (for direct production)—electric energy, gas, fuel, compressed air, water treatment plant, internal communications, internal transportation, sewage, tailings treatment plant, etc. If applicable, flow diagrams and material and/or energy balances must be provided. A global estimate of capital expenditures (as a function of plant size, if applicable) should also be included.

Service facilities. These include plant security, medical facilities, dining room, and other personnel-related facilities. A global estimate of capital expenditures (as a function of plant size, if applicable) is required.

Raw materials and supplies. For each production level, an estimate must be made of quantity, quality requirements, annual consumption, availability, unit price, and consumption per unit of product. For electric energy, it is customary to indicate installed power, processes at constant and at variable load, and maximum illumination-related load. An estimate of size and value must be given for all raw materials and for in-production or finished-product inventories expected for the process.

Transportation expenses—for raw materials, fuels, and intermediate and finished products. If contracts are to be awarded to third parties, their availability and maximum requirements (at start-up and in the long run), as well as prices, must be included.

Manpower requirements. Estimate of personnel required at different levels in each unit, including labor, direct supervisory personnel, indirect supervisory personnel, service people (labs, maintenance, security), administration, marketing, and management. Detail wage and salary structures, as well as fringe benefit charges. Any contractual obligations should also be included. Any need for specially trained personnel, training facilities, start-up consultants, etc., should also be reported.

The project engineering phase of the analysis should produce a summary bar diagram describing target completion date for all major project components, as well as a plan for project implementation until "normal" operation is achieved. Such a bar chart is illustrated in Fig. 12-2.

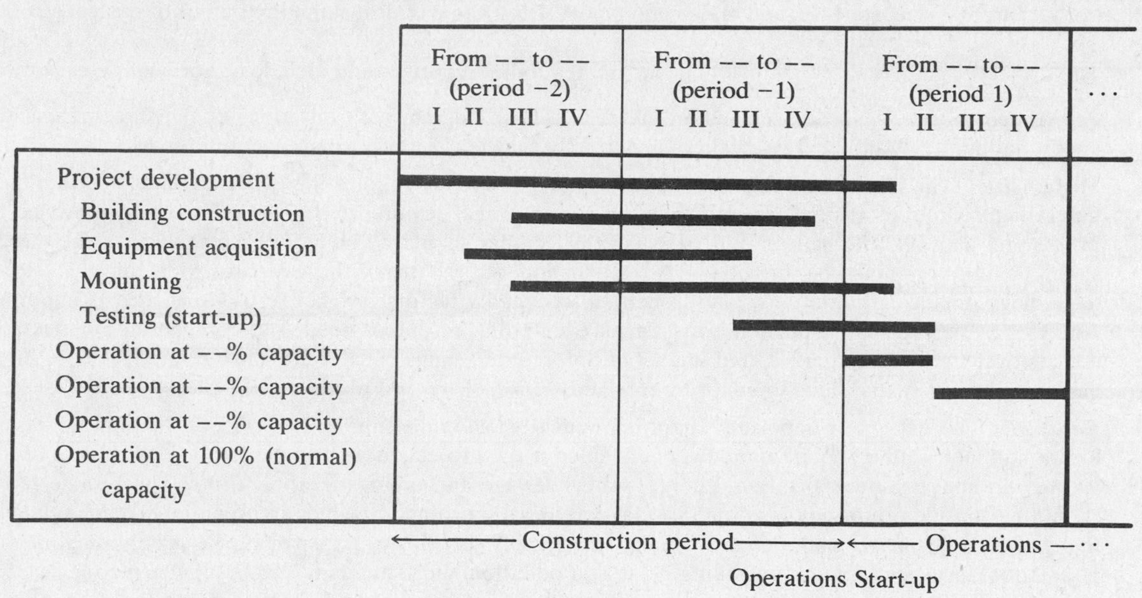

Fig. 12-2

12.5 COST ESTIMATION

This section of the economic feasibility report falls into two main parts: the analysis of *investment costs* (or *first costs*) and the analysis of *operating costs.*

Example 12.6 Itemize the components of the investment costs for a typical industrial project.

The investment costs include all expenditures related to the implementation of the project, from its conception until start-up. A distinction is usually made between *depreciable* and *nondepreciable* investment costs.

Fixed capital costs. (Depreciable, except for land purchasing). These include costs of: residential buildings (offices, cafeteria, etc.); industrial buildings (including warehouses); roads, energy lines, railroads, and other infrastructure; hauling, loading, and unloading equipment; industrial machinery and equipment; spare parts (includes maintenance, repair, and standby equipment); ancillary facilities (electrical substation, laboratories, transformers, fire extinguishers, etc.); office furniture. Usually only buildings and main equipment and facilities are estimated accurately. Spare parts, furniture, etc., are often estimated as a percentage of those items.

Operating capital costs. This is the amount of money required to start up the project and keep it working. The operating capital costs usually increase until the project reaches the level of normal operation; then they stay on a plateau throughout the project's lifetime, and are recovered in the final year of operation. They cannot be depreciated. They include: cash (to pay salaries, to cover emergencies, and—sometimes—to help in operating process); circulating capital (accounts receivable minus accounts payable); stocks and inventories (general merchandise, finished or intermediate or secondary products, raw materials, in-transit material, packages, consumption materials, etc.); material handling (loading and transportation and unloading from and to the warehouses, cost of inventory control and insurance, protection of inventories, etc.).

The analysis of first cost should produce a table such as Table 12-1 (for each alternative size of project considered).

Table 12-1

Item	**$**
Feasibility studies	
Land	
Land preparation	
Construction & installation (buildings)	
Equipment (incl. insurance, transportation, installation & start-up)	
For the main plant	
For ancillary facilities	
For construction	
Patents and royalties	
Supervision	
Consultantships (legal and engineering)	
Start-up costs	
Subtotal	
Contingency costs (% of subtotal)	
Operating capital	
TOTAL INVESTMENT COST	$

Typical errors in the analysis of investment costs are: (i) underestimation of transportation, installation, and start-up costs; (ii) underestimation of time needed to construct the project; (iii) underestimation of the operating capital; (iv) underestimation of the time needed to test-run equipment and to reach the level of normal operation; and (most common) (v) omission of a sensitivity study of project size. Keep in mind that first costs, being incurred in year 0, are not discounted. Thus, too-large estimates of investment costs can kill a project's feasibility a lot faster than any overestimation of operating expenses.

Example 12.7 Itemize the components of the operating costs for a typical industrial project.

Direct costs: raw material (includes cost of handling); direct materials (explosives, catalysts, grinding balls, packages, etc.); direct labor (including direct supervision)—salaries, fringe benefits, overtime, etc.; direct-production utilities (energy, fuel, lubricants, steam, water, etc.).

Indirect costs: indirect labor (salaries, fringe benefits, overtime, etc., for general supervision, maintenance, general engineering, security, plant protection, quality control, laboratories); indirect materials (e.g., lab reagents); other indirect costs (health clinic, recreation and eating facilities, transportation of personnel, communications, lights, cleaning, etc.); employee benefits (child-care center, gymnasium, etc.).

Overhead costs: administrative costs (salaries of managerial personnel, secretaries, legal and engineering staff; rent, office cleaning, office materials, reproduction, etc.); fixed charges (taxes, insurance); selling expenses (salesmen, commissions, travel, market surveys, entertainment of clients, displays, sales space, etc.); research and development; financing charges (interest and loan payments); bad debts; contributions (not in excess of 5% of taxable income); losses by fire, theft, etc., not covered by insurance.

The analysis of the operating costs should produce a table such as Table 12-2 (for each alternative size of project considered).

Table 12-2

Item	*Year 1*	*Year 2* ··· *Year n*
Direct costs		
Labor		
Material		
Indirect costs		
Labor		
Material		
Other		
Overhead costs		
Administrative costs		
Selling costs		
Other		
TOTAL OPERATING COST ($)		

The most common error in the analysis of operating costs is the wrong estimation of plant utilization (i.e., designing with overcapacity), which strongly affects direct costs.

12.6 ESTIMATION OF REVENUES

A projection of annual income has to be made, with consideration of: (i) sales of main and secondary products; (ii) services provided; (iii) recuperation of operating capital (typically occurs at the end of year n, when stocks and inventories are depleted, circulating capital is settled, and material handling stops).

The most common error in this phase of the analysis is to assume that all the production will be sold. Sometimes market conditions will not permit this to happen; or the product quality may fluctuate, and sizable quantities may have to be recycled or scrapped.

Another common error, from a company point of view, is to forget that revenue consists only in actual *incremental* receipts. For example, if an ice cream distributor is evaluating a new line of products, actual net receipts attributable to this product is given by the expected sales *less* the loss in sales of current products because of customers' shifting preferences. In the extreme case, the public may not be spending more money for the company's products (although the new product is experiencing brisk sales); rather, they have ceased buying some of the old products and are using that money to buy the new one!

12.7 FINANCING

Once the cost and revenue analyses are ready, (*11.15*) and (*11.19*) may be used to prepare a summary table of yearly cash flows (see Table 12-3). Then a specific performance indicator, such as after-tax PW or after-tax ROR, can be computed. In fact, it is recommended that both these indicators be calculated. For, on the one hand, if the company has a good estimate of its (after-tax) MARR, then the PW corresponding to that MARR will represent to the company its net increase in assets as a result of the project once the initial investment (and all operating costs) have been recovered and the MARR realized. On the other hand, financial institutions prefer the ROR, since it is independent of the company's MARR and is useful in choosing among independent projects that are similar in size and duration (see Section 9.2).

Table 12-3

(1)	Sales
(2)	Services provided
(3)	Recovered operating capital
(4)	TOTAL RECEIPTS
(5)	Less: Direct labor
(6)	Less: Direct material
(7)	Less: Maintenance
(8)	Less: Indirect material
(9)	Less: Indirect labor
(10)	Less: Overhead
(11)	Less: Other expenses
(12)	Net income before taxes
(13)	Less: Depreciation
(14)	Net taxable income
(15)	Less: Taxes [rate × (14)]
(16)	After-tax net income
(17)	Depreciation
(18)	After-tax net cash flow

It should be noted that, generally, not one but two economic evaluations must be presented to a lending institution. One of them analyzes the project as if it were going to be completely financed by the owner. The results of this analysis reflect the project's potential of success (and its ability to generate enough revenue to pay the loan). The second analysis takes account of the size of the loan and of the corresponding schedule of payments. This analysis furnishes the owner with a better representation of the economic potential of the project as it is actually intended to be financed. Both sorts of evaluation are carried out using the techniques of Chapter 11—see, in particular, Problems 11.19 and 11.21. (That Chapter 11 dealt with *level* series of before-tax cash flows is obviously inessential.)

Appendix A

Compound Interest Factors—Annual Compounding

COMPOUND INTEREST FACTORS - ANNUAL COMPOUNDING

INTEREST RATE = 0.25 PERCENT

N	SINGLE-PAYMENT COMPOUND-AMOUNT FACTOR (F/P)	UNIFORM-SERIES COMPOUND-AMOUNT FACTOR (F/A)	UNIFORM-SERIES CAPITAL-RECOVERY FACTOR (A/P)	GRADIENT SERIES FACTOR (A/G)
1	1.0025	1.0000	1.00250	.0000
2	1.0050	2.0025	.50188	.4994
3	1.0075	3.0075	.33500	.9983
4	1.0100	4.0150	.25156	1.4969
5	1.0126	5.0251	.20150	1.9950
6	1.0151	6.0376	.16813	2.4927
7	1.0176	7.0527	.14429	2.9900
8	1.0202	8.0704	.12641	3.4869
9	1.0227	9.0905	.11250	3.9834
10	1.0253	10.1133	.10138	4.4794
11	1.0278	11.1385	.09228	4.9750
12	1.0304	12.1664	.08469	5.4702
13	1.0330	13.1968	.07828	5.9650
14	1.0356	14.2298	.07278	6.4594
15	1.0382	15.2654	.06801	6.9534
16	1.0408	16.3035	.06384	7.4469
17	1.0434	17.3443	.06016	7.9401
18	1.0460	18.3876	.05688	8.4328
19	1.0486	19.4336	.05396	8.9251
20	1.0512	20.4822	.05132	9.4170
21	1.0538	21.5334	.04894	9.9085
22	1.0565	22.5872	.04677	10.3995
23	1.0591	23.6437	.04479	10.8901
24	1.0618	24.7028	.04298	11.3804
25	1.0644	25.7646	.04131	11.8702
26	1.0671	26.8290	.03977	12.3596
27	1.0697	27.8961	.03835	12.8485
28	1.0724	28.9658	.03702	13.3371
29	1.0751	30.0382	.03579	13.8252
30	1.0778	31.1133	.03464	14.3130
31	1.0805	32.1911	.03356	14.8003
32	1.0832	33.2716	.03256	15.2872
33	1.0859	34.3547	.03161	15.7736
34	1.0886	35.4406	.03072	16.2597
35	1.0913	36.5292	.02988	16.7454
40	1.1050	42.0132	.02630	19.1673
45	1.1189	47.5661	.02352	21.5789
50	1.1330	53.1887	.02130	23.9802
55	1.1472	58.8819	.01948	26.3710
60	1.1616	64.6467	.01797	28.7514
65	1.1762	70.4839	.01669	31.1215
70	1.1910	76.3944	.01559	33.4812
75	1.2059	82.3792	.01464	35.8305
80	1.2211	88.4392	.01381	38.1694
85	1.2364	94.5753	.01307	40.4980
90	1.2520	100.7885	.01242	42.8162
95	1.2677	107.0797	.01184	45.1241
100	1.2836	113.4500	.01131	47.4216

COMPOUND INTEREST FACTORS - ANNUAL COMPOUNDING

INTEREST RATE = 0.50 PERCENT

N	SINGLE-PAYMENT COMPOUND-AMOUNT FACTOR (F/P)	UNIFORM-SERIES COMPOUND-AMOUNT FACTOR (F/A)	UNIFORM-SERIES CAPITAL-RECOVERY FACTOR (A/P)	GRADIENT SERIES FACTOR (A/G)
1	1.0050	1.0000	1.00500	.0000
2	1.0100	2.0050	.50375	.4988
3	1.0151	3.0150	.33667	.9967
4	1.0202	4.0301	.25313	1.4938
5	1.0253	5.0503	.20301	1.9900
6	1.0304	6.0755	.16960	2.4855
7	1.0355	7.1059	.14573	2.9801
8	1.0407	8.1414	.12783	3.4738
9	1.0459	9.1821	.11391	3.9668
10	1.0511	10.2280	.10277	4.4589
11	1.0564	11.2792	.09366	4.9501
12	1.0617	12.3356	.08607	5.4406
13	1.0670	13.3972	.07964	5.9302
14	1.0723	14.4642	.07414	6.4190
15	1.0777	15.5365	.06936	6.9069
16	1.0831	16.6142	.06519	7.3940
17	1.0885	17.6973	.06151	7.8803
18	1.0939	18.7858	.05823	8.3658
19	1.0994	19.8797	.05530	8.8504
20	1.1049	20.9791	.05267	9.3342
21	1.1104	22.0840	.05028	9.8172
22	1.1160	23.1944	.04811	10.2993
23	1.1216	24.3104	.04613	10.7806
24	1.1272	25.4320	.04432	11.2611
25	1.1328	26.5591	.04265	11.7407
26	1.1385	27.6919	.04111	12.2195
27	1.1442	28.8304	.03969	12.6975
28	1.1499	29.9745	.03836	13.1747
29	1.1556	31.1244	.03713	13.6510
30	1.1614	32.2800	.03598	14.1265
31	1.1672	33.4414	.03490	14.6012
32	1.1730	34.6086	.03389	15.0750
33	1.1789	35.7817	.03295	15.5480
34	1.1848	36.9606	.03206	16.0202
35	1.1907	38.1454	.03122	16.4915
40	1.2208	44.1588	.02765	18.8359
45	1.2516	50.3242	.02487	21.1595
50	1.2832	56.6452	.02265	23.4624
55	1.3156	63.1258	.02084	25.7447
60	1.3489	69.7700	.01933	28.0064
65	1.3829	76.5821	.01806	30.2475
70	1.4178	83.5661	.01697	32.4680
75	1.4536	90.7265	.01602	34.6679
80	1.4903	98.0677	.01520	36.8474
85	1.5280	105.5943	.01447	39.0065
90	1.5666	113.3109	.01383	41.1451
95	1.6061	121.2224	.01325	43.2633
100	1.6467	129.3337	.01273	45.3613

COMPOUND INTEREST FACTORS - ANNUAL COMPOUNDING

INTEREST RATE = 0.75 PERCENT

N	SINGLE-PAYMENT COMPOUND-AMOUNT FACTOR (F/P)	UNIFORM-SERIES COMPOUND-AMOUNT FACTOR (F/A)	UNIFORM-SERIES CAPITAL-RECOVERY FACTOR (A/P)	GRADIENT SERIES FACTOR (A/G)
1	1.0075	1.0000	1.00750	.0000
2	1.0151	2.0075	.50563	.4981
3	1.0227	3.0226	.33835	.9950
4	1.0303	4.0452	.25471	1.4907
5	1.0381	5.0756	.20452	1.9851
6	1.0459	6.1136	.17107	2.4782
7	1.0537	7.1595	.14717	2.9701
8	1.0616	8.2132	.12926	3.4608
9	1.0696	9.2748	.11532	3.9502
10	1.0776	10.3443	.10417	4.4384
11	1.0857	11.4219	.09505	4.9253
12	1.0938	12.5076	.08745	5.4110
13	1.1020	13.6014	.08102	5.8954
14	1.1103	14.7034	.07551	6.3786
15	1.1186	15.8137	.07074	6.8606
16	1.1270	16.9323	.06656	7.3413
17	1.1354	18.0593	.06287	7.8207
18	1.1440	19.1947	.05960	8.2989
19	1.1525	20.3387	.05667	8.7759
20	1.1612	21.4912	.05403	9.2516
21	1.1699	22.6524	.05165	9.7261
22	1.1787	23.8223	.04948	10.1994
23	1.1875	25.0010	.04750	10.6714
24	1.1964	26.1885	.04568	11.1422
25	1.2054	27.3849	.04402	11.6117
26	1.2144	28.5903	.04248	12.0800
27	1.2235	29.8047	.04105	12.5470
28	1.2327	31.0282	.03973	13.0128
29	1.2420	32.2609	.03850	13.4774
30	1.2513	33.5029	.03735	13.9407
31	1.2607	34.7542	.03627	14.4028
32	1.2701	36.0148	.03527	14.8636
33	1.2796	37.2849	.03432	15.3232
34	1.2892	38.5646	.03343	15.7816
35	1.2989	39.8538	.03259	16.2387
40	1.3483	46.4465	.02903	18.5058
45	1.3997	53.2901	.02627	20.7421
50	1.4530	60.3943	.02406	22.9476
55	1.5083	67.7688	.02226	25.1223
60	1.5657	75.4241	.02076	27.2665
65	1.6253	83.3709	.01949	29.3801
70	1.6872	91.6201	.01841	31.4634
75	1.7514	100.1833	.01748	33.5163
80	1.8180	109.0725	.01667	35.5391
85	1.8873	118.3001	.01595	37.5318
90	1.9591	127.8790	.01532	39.4946
95	2.0337	137.8225	.01476	41.4277
100	2.1111	148.1445	.01425	43.3311

COMPOUND INTEREST FACTORS - ANNUAL COMPOUNDING

INTEREST RATE = 1.00 PERCENT

N	SINGLE-PAYMENT COMPOUND-AMOUNT FACTOR (F/P)	UNIFORM-SERIES COMPOUND-AMOUNT FACTOR (F/A)	UNIFORM-SERIES CAPITAL-RECOVERY FACTOR (A/P)	GRADIENT SERIES FACTOR (A/G)
1	1.0100	1.0000	1.01000	.0000
2	1.0201	2.0100	.50751	.4975
3	1.0303	3.0301	.34002	.9934
4	1.0406	4.0604	.25628	1.4876
5	1.0510	5.1010	.20604	1.9801
6	1.0615	6.1520	.17255	2.4710
7	1.0721	7.2135	.14863	2.9602
8	1.0829	8.2857	.13069	3.4478
9	1.0937	9.3685	.11674	3.9337
10	1.1046	10.4622	.10558	4.4179
11	1.1157	11.5668	.09645	4.9005
12	1.1268	12.6825	.08885	5.3815
13	1.1381	13.8093	.08241	5.8607
14	1.1495	14.9474	.07690	6.3384
15	1.1610	16.0969	.07212	6.8143
16	1.1726	17.2579	.06794	7.2886
17	1.1843	18.4304	.06426	7.7613
18	1.1961	19.6147	.06098	8.2323
19	1.2081	20.8109	.05805	8.7017
20	1.2202	22.0190	.05542	9.1694
21	1.2324	23.2392	.05303	9.6354
22	1.2447	24.4716	.05086	10.0998
23	1.2572	25.7163	.04889	10.5626
24	1.2697	26.9735	.04707	11.0237
25	1.2824	28.2432	.04541	11.4831
26	1.2953	29.5256	.04387	11.9409
27	1.3082	30.8209	.04245	12.3971
28	1.3213	32.1291	.04112	12.8516
29	1.3345	33.4504	.03990	13.3044
30	1.3478	34.7849	.03875	13.7557
31	1.3613	36.1327	.03768	14.2052
32	1.3749	37.4941	.03667	14.6532
33	1.3887	38.8690	.03573	15.0995
34	1.4026	40.2577	.03484	15.5441
35	1.4166	41.6603	.03400	15.9871
40	1.4889	48.8864	.03046	18.1776
45	1.5648	56.4811	.02771	20.3273
50	1.6446	64.4632	.02551	22.4363
55	1.7285	72.8525	.02373	24.5049
60	1.8167	81.6697	.02224	26.5333
65	1.9094	90.9366	.02100	28.5217
70	2.0068	100.6763	.01993	30.4703
75	2.1091	110.9128	.01902	32.3793
80	2.2167	121.6715	.01822	34.2492
85	2.3298	132.9790	.01752	36.0801
90	2.4486	144.8633	.01690	37.8724
95	2.5735	157.3538	.01636	39.6265
100	2.7048	170.4814	.01587	41.3426

COMPOUND INTEREST FACTORS - ANNUAL COMPOUNDING

INTEREST RATE = 1.25 PERCENT

N	SINGLE-PAYMENT COMPOUND-AMOUNT FACTOR (F/P)	UNIFORM-SERIES COMPOUND-AMOUNT FACTOR (F/A)	UNIFORM-SERIES CAPITAL-RECOVERY FACTOR (A/P)	GRADIENT SERIES FACTOR (A/G)
1	1.0125	1.0000	1.01250	.0000
2	1.0252	2.0125	.50939	.4969
3	1.0380	3.0377	.34170	.9917
4	1.0509	4.0756	.25786	1.4845
5	1.0641	5.1266	.20756	1.9752
6	1.0774	6.1907	.17403	2.4638
7	1.0909	7.2680	.15009	2.9503
8	1.1045	8.3589	.13213	3.4348
9	1.1183	9.4634	.11817	3.9172
10	1.1323	10.5817	.10700	4.3975
11	1.1464	11.7139	.09787	4.8758
12	1.1608	12.8604	.09026	5.3520
13	1.1753	14.0211	.08382	5.8262
14	1.1900	15.1964	.07831	6.2982
15	1.2048	16.3863	.07353	6.7682
16	1.2199	17.5912	.06935	7.2362
17	1.2351	18.8111	.06566	7.7021
18	1.2506	20.0462	.06238	8.1659
19	1.2662	21.2968	.05946	8.6277
20	1.2820	22.5630	.05682	9.0874
21	1.2981	23.8450	.05444	9.5450
22	1.3143	25.1431	.05227	10.0006
23	1.3307	26.4574	.05030	10.4542
24	1.3474	27.7881	.04849	10.9056
25	1.3642	29.1354	.04682	11.3551
26	1.3812	30.4996	.04529	11.8024
27	1.3985	31.8809	.04387	12.2478
28	1.4160	33.2794	.04255	12.6911
29	1.4337	34.6954	.04132	13.1323
30	1.4516	36.1291	.04018	13.5715
31	1.4698	37.5807	.03911	14.0086
32	1.4881	39.0504	.03811	14.4438
33	1.5067	40.5386	.03717	14.8768
34	1.5256	42.0453	.03628	15.3079
35	1.5446	43.5709	.03545	15.7369
40	1.6436	51.4896	.03192	17.8515
45	1.7489	59.9157	.02919	19.9156
50	1.8610	68.8818	.02702	21.9295
55	1.9803	78.4225	.02525	23.8936
60	2.1072	88.5745	.02379	25.8083
65	2.2422	99.3771	.02256	27.6741
70	2.3859	110.8720	.02152	29.4913
75	2.5388	123.1035	.02062	31.2605
80	2.7015	136.1188	.01985	32.9822
85	2.8746	149.9682	.01917	34.6570
90	3.0588	164.7050	.01857	36.2855
95	3.2548	180.3862	.01804	37.8682
100	3.4634	197.0723	.01757	39.4058

COMPOUND INTEREST FACTORS - ANNUAL COMPOUNDING

INTEREST RATE = 1.50 PERCENT

N	SINGLE-PAYMENT COMPOUND-AMOUNT FACTOR (F/P)	UNIFORM-SERIES COMPOUND-AMOUNT FACTOR (F/A)	UNIFORM-SERIES CAPITAL-RECOVERY FACTOR (A/P)	GRADIENT SERIES FACTOR (A/G)
1	1.0150	1.0000	1.01500	.0000
2	1.0302	2.0150	.51128	.4963
3	1.0457	3.0452	.34338	.9901
4	1.0614	4.0909	.25944	1.4814
5	1.0773	5.1523	.20909	1.9702
6	1.0934	6.2296	.17553	2.4566
7	1.1098	7.3230	.15156	2.9405
8	1.1265	8.4328	.13358	3.4219
9	1.1434	9.5593	.11961	3.9008
10	1.1605	10.7027	.10843	4.3772
11	1.1779	11.8633	.09929	4.8512
12	1.1956	13.0412	.09168	5.3227
13	1.2136	14.2368	.08524	5.7917
14	1.2318	15.4504	.07972	6.2582
15	1.2502	16.6821	.07494	6.7223
16	1.2690	17.9324	.07077	7.1839
17	1.2880	19.2014	.06708	7.6431
18	1.3073	20.4894	.06381	8.0997
19	1.3270	21.7967	.06088	8.5539
20	1.3469	23.1237	.05825	9.0057
21	1.3671	24.4705	.05587	9.4550
22	1.3876	25.8376	.05370	9.9018
23	1.4084	27.2251	.05173	10.3462
24	1.4295	28.6335	.04992	10.7881
25	1.4509	30.0630	.04826	11.2276
26	1.4727	31.5140	.04673	11.6646
27	1.4948	32.9867	.04532	12.0992
28	1.5172	34.4815	.04400	12.5313
29	1.5400	35.9987	.04278	12.9610
30	1.5631	37.5387	.04164	13.3883
31	1.5865	39.1018	.04057	13.8131
32	1.6103	40.6883	.03958	14.2355
33	1.6345	42.2986	.03864	14.6555
34	1.6590	43.9331	.03776	15.0731
35	1.6839	45.5921	.03693	15.4882
40	1.8140	54.2679	.03343	17.5277
45	1.9542	63.6142	.03072	19.5074
50	2.1052	73.6828	.02857	21.4277
55	2.2679	84.5296	.02683	23.2894
60	2.4432	96.2147	.02539	25.0930
65	2.6320	108.8028	.02419	26.8393
70	2.8355	122.3638	.02317	28.5290
75	3.0546	136.9728	.02230	30.1631
80	3.2907	152.7109	.02155	31.7423
85	3.5450	169.6652	.02089	33.2676
90	3.8189	187.9299	.02032	34.7399
95	4.1141	207.6061	.01982	36.1602
100	4.4320	228.8030	.01937	37.5295

COMPOUND INTEREST FACTORS - ANNUAL COMPOUNDING

INTEREST RATE = 2.00 PERCENT

N	SINGLE-PAYMENT COMPOUND-AMOUNT FACTOR (F/P)	UNIFORM-SERIES COMPOUND-AMOUNT FACTOR (F/A)	UNIFORM-SERIES CAPITAL-RECOVERY FACTOR (A/P)	GRADIENT SERIES FACTOR (A/G)
1	1.0200	1.0000	1.02000	.0000
2	1.0404	2.0200	.51505	.4950
3	1.0612	3.0604	.34675	.9868
4	1.0824	4.1216	.26262	1.4752
5	1.1041	5.2040	.21216	1.9604
6	1.1262	6.3081	.17853	2.4423
7	1.1487	7.4343	.15451	2.9208
8	1.1717	8.5830	.13651	3.3961
9	1.1951	9.7546	.12252	3.8681
10	1.2190	10.9497	.11133	4.3367
11	1.2434	12.1687	.10218	4.8021
12	1.2682	13.4121	.09456	5.2642
13	1.2936	14.6803	.08812	5.7231
14	1.3195	15.9739	.08260	6.1786
15	1.3459	17.2934	.07783	6.6309
16	1.3728	18.6393	.07365	7.0799
17	1.4002	20.0121	.06997	7.5256
18	1.4282	21.4123	.06670	7.9681
19	1.4568	22.8406	.06378	8.4073
20	1.4859	24.2974	.06116	8.8433
21	1.5157	25.7833	.05878	9.2760
22	1.5460	27.2990	.05663	9.7055
23	1.5769	28.8450	.05467	10.1317
24	1.6084	30.4219	.05287	10.5547
25	1.6406	32.0303	.05122	10.9745
26	1.6734	33.6709	.04970	11.3910
27	1.7069	35.3443	.04829	11.8043
28	1.7410	37.0512	.04699	12.2145
29	1.7758	38.7922	.04578	12.6214
30	1.8114	40.5681	.04465	13.0251
31	1.8476	42.3794	.04360	13.4257
32	1.8845	44.2270	.04261	13.8230
33	1.9222	46.1116	.04169	14.2172
34	1.9607	48.0338	.04082	14.6083
35	1.9999	49.9945	.04000	14.9961
40	2.2080	60.4020	.03656	16.8885
45	2.4379	71.8927	.03391	18.7034
50	2.6916	84.5794	.03182	20.4420
55	2.9717	98.5865	.03014	22.1057
60	3.2810	114.0515	.02877	23.6961
65	3.6225	131.1262	.02763	25.2147
70	3.9996	149.9779	.02667	26.6632
75	4.4158	170.7918	.02586	28.0434
80	4.8754	193.7720	.02516	29.3572
85	5.3829	219.1439	.02456	30.6064
90	5.9431	247.1567	.02405	31.7929
95	6.5617	278.0850	.02360	32.9189
100	7.2446	312.2323	.02320	33.9863

COMPOUND INTEREST FACTORS - ANNUAL COMPOUNDING

INTEREST RATE = 3.00 PERCENT

N	SINGLE-PAYMENT COMPOUND-AMOUNT FACTOR (F/P)	UNIFORM-SERIES COMPOUND-AMOUNT FACTOR (F/A)	UNIFORM-SERIES CAPITAL-RECOVERY FACTOR (A/P)	GRADIENT SERIES FACTOR (A/G)
1	1.0300	1.0000	1.03000	.0000
2	1.0609	2.0300	.52261	.4926
3	1.0927	3.0909	.35353	.9803
4	1.1255	4.1836	.26903	1.4631
5	1.1593	5.3091	.21835	1.9409
6	1.1941	6.4684	.18460	2.4138
7	1.2299	7.6625	.16051	2.8819
8	1.2668	8.8923	.14246	3.3450
9	1.3048	10.1591	.12843	3.8032
10	1.3439	11.4639	.11723	4.2565
11	1.3842	12.8078	.10808	4.7049
12	1.4258	14.1920	.10046	5.1485
13	1.4685	15.6178	.09403	5.5872
14	1.5126	17.0863	.08853	6.0210
15	1.5580	18.5989	.08377	6.4500
16	1.6047	20.1569	.07961	6.8742
17	1.6528	21.7616	.07595	7.2936
18	1.7024	23.4144	.07271	7.7081
19	1.7535	25.1169	.06981	8.1179
20	1.8061	26.8704	.06722	8.5229
21	1.8603	28.6765	.06487	8.9231
22	1.9161	30.5368	.06275	9.3186
23	1.9736	32.4529	.06081	9.7093
24	2.0328	34.4265	.05905	10.0954
25	2.0938	36.4593	.05743	10.4768
26	2.1566	38.5530	.05594	10.8535
27	2.2213	40.7096	.05456	11.2255
28	2.2879	42.9309	.05329	11.5930
29	2.3566	45.2189	.05211	11.9558
30	2.4273	47.5754	.05102	12.3141
31	2.5001	50.0027	.05000	12.6678
32	2.5751	52.5028	.04905	13.0169
33	2.6523	55.0778	.04816	13.3616
34	2.7319	57.7302	.04732	13.7018
35	2.8139	60.4621	.04654	14.0375
40	3.2620	75.4013	.04326	15.6502
45	3.7816	92.7199	.04079	17.1556
50	4.3839	112.7969	.03887	18.5575
55	5.0821	136.0716	.03735	19.8600
60	5.8916	163.0534	.03613	21.0674
65	6.8300	194.3328	.03515	22.1841
70	7.9178	230.5941	.03434	23.2145
75	9.1789	272.6309	.03367	24.1634
80	10.6409	321.3630	.03311	25.0353
85	12.3357	377.8570	.03265	25.8349
90	14.3005	443.3489	.03226	26.5667
95	16.5782	519.2720	.03193	27.2351
100	19.2186	607.2877	.03165	27.8444

COMPOUND INTEREST FACTORS - ANNUAL COMPOUNDING

INTEREST RATE = 4.00 PERCENT

N	SINGLE-PAYMENT COMPOUND-AMOUNT FACTOR (F/P)	UNIFORM-SERIES COMPOUND-AMOUNT FACTOR (F/A)	UNIFORM-SERIES CAPITAL-RECOVERY FACTOR (A/P)	GRADIENT SERIES FACTOR (A/G)
1	1.0400	1.0000	1.04000	.0000
2	1.0816	2.0400	.53020	.4902
3	1.1249	3.1216	.36035	.9739
4	1.1699	4.2465	.27549	1.4510
5	1.2167	5.4163	.22463	1.9216
6	1.2653	6.6330	.19076	2.3857
7	1.3159	7.8983	.16661	2.8433
8	1.3686	9.2142	.14853	3.2944
9	1.4233	10.5828	.13449	3.7391
10	1.4802	12.0061	.12329	4.1773
11	1.5395	13.4864	.11415	4.6090
12	1.6010	15.0258	.10655	5.0343
13	1.6651	16.6268	.10014	5.4533
14	1.7317	18.2919	.09467	5.8659
15	1.8009	20.0236	.08994	6.2721
16	1.8730	21.8245	.08582	6.6720
17	1.9479	23.6975	.08220	7.0656
18	2.0258	25.6454	.07899	7.4530
19	2.1068	27.6712	.07614	7.8342
20	2.1911	29.7781	.07358	8.2091
21	2.2788	31.9692	.07128	8.5779
22	2.3699	34.2480	.06920	8.9407
23	2.4647	36.6179	.06731	9.2973
24	2.5633	39.0826	.06559	9.6479
25	2.6658	41.6459	.06401	9.9925
26	2.7725	44.3117	.06257	10.3312
27	2.8834	47.0842	.06124	10.6640
28	2.9987	49.9676	.06001	10.9909
29	3.1187	52.9663	.05888	11.3120
30	3.2434	56.0849	.05783	11.6274
31	3.3731	59.3283	.05686	11.9371
32	3.5081	62.7015	.05595	12.2411
33	3.6484	66.2095	.05510	12.5396
34	3.7943	69.8579	.05431	12.8324
35	3.9461	73.6522	.05358	13.1198
40	4.8010	95.0255	.05052	14.4765
45	5.8412	121.0294	.04826	15.7047
50	7.1067	152.6671	.04655	16.8122
55	8.6464	191.1592	.04523	17.8070
60	10.5196	237.9907	.04420	18.6972
65	12.7987	294.9684	.04339	19.4909
70	15.5716	364.2905	.04275	20.1961
75	18.9453	448.6314	.04223	20.8206
80	23.0498	551.2450	.04181	21.3718
85	28.0436	676.0901	.04148	21.8569
90	34.1193	827.9833	.04121	22.2826
95	41.5114	1012.7846	.04099	22.6550
100	50.5049	1237.6237	.04081	22.9800

COMPOUND INTEREST FACTORS - ANNUAL COMPOUNDING

INTEREST RATE = 5.00 PERCENT

N	SINGLE-PAYMENT COMPOUND-AMOUNT FACTOR (F/P)	UNIFORM-SERIES COMPOUND-AMOUNT FACTOR (F/A)	UNIFORM-SERIES CAPITAL-RECOVERY FACTOR (A/P)	GRADIENT SERIES FACTOR (A/G)
1	1.0500	1.0000	1.05000	.0000
2	1.1025	2.0500	.53780	.4878
3	1.1576	3.1525	.36721	.9675
4	1.2155	4.3101	.28201	1.4391
5	1.2763	5.5256	.23097	1.9025
6	1.3401	6.8019	.19702	2.3579
7	1.4071	8.1420	.17282	2.8052
8	1.4775	9.5491	.15472	3.2445
9	1.5513	11.0266	.14069	3.6758
10	1.6289	12.5779	.12950	4.0991
11	1.7103	14.2068	.12039	4.5144
12	1.7959	15.9171	.11283	4.9219
13	1.8856	17.7130	.10646	5.3215
14	1.9799	19.5986	.10102	5.7133
15	2.0789	21.5786	.09634	6.0973
16	2.1829	23.6575	.09227	6.4736
17	2.2920	25.8404	.08870	6.8423
18	2.4066	28.1324	.08555	7.2034
19	2.5270	30.5390	.08275	7.5569
20	2.6533	33.0660	.08024	7.9030
21	2.7860	35.7193	.07800	8.2416
22	2.9253	38.5052	.07597	8.5730
23	3.0715	41.4305	.07414	8.8971
24	3.2251	44.5020	.07247	9.2140
25	3.3864	47.7271	.07095	9.5238
26	3.5557	51.1135	.06956	9.8266
27	3.7335	54.6691	.06829	10.1224
28	3.9201	58.4026	.06712	10.4114
29	4.1161	62.3227	.06605	10.6936
30	4.3219	66.4388	.06505	10.9691
31	4.5380	70.7608	.06413	11.2381
32	4.7649	75.2988	.06328	11.5005
33	5.0032	80.0638	.06249	11.7566
34	5.2533	85.0670	.06176	12.0063
35	5.5160	90.3203	.06107	12.2498
40	7.0400	120.7998	.05828	13.3775
45	8.9850	159.7002	.05626	14.3644
50	11.4674	209.3480	.05478	15.2233
55	14.6356	272.7126	.05367	15.9664
60	18.6792	353.5837	.05283	16.6062
65	23.8399	456.7980	.05219	17.1541
70	30.4264	588.5285	.05170	17.6212
75	38.8327	756.6537	.05132	18.0176
80	49.5614	971.2288	.05103	18.3526
85	63.2544	1245.0871	.05080	18.6346
90	80.7304	1594.6073	.05063	18.8712
95	103.0347	2040.6935	.05049	19.0689
100	131.5013	2610.0252	.05038	19.2337

COMPOUND INTEREST FACTORS - ANNUAL COMPOUNDING

INTEREST RATE = 6.00 PERCENT

N	SINGLE-PAYMENT COMPOUND-AMOUNT FACTOR (F/P)	UNIFORM-SERIES COMPOUND-AMOUNT FACTOR (F/A)	UNIFORM-SERIES CAPITAL-RECOVERY FACTOR (A/P)	GRADIENT SERIES FACTOR (A/G)
1	1.0600	1.0000	1.06000	.0000
2	1.1236	2.0600	.54544	.4854
3	1.1910	3.1836	.37411	.9612
4	1.2625	4.3746	.28859	1.4272
5	1.3382	5.6371	.23740	1.8836
6	1.4185	6.9753	.20336	2.3304
7	1.5036	8.3938	.17914	2.7676
8	1.5938	9.8975	.16104	3.1952
9	1.6895	11.4913	.14702	3.6133
10	1.7908	13.1808	.13587	4.0220
11	1.8983	14.9716	.12679	4.4213
12	2.0122	16.8699	.11928	4.8113
13	2.1329	18.8821	.11296	5.1920
14	2.2609	21.0151	.10758	5.5635
15	2.3966	23.2760	.10296	5.9260
16	2.5404	25.6725	.09895	6.2794
17	2.6928	28.2129	.09544	6.6240
18	2.8543	30.9057	.09236	6.9597
19	3.0256	33.7600	.08962	7.2867
20	3.2071	36.7856	.08718	7.6051
21	3.3996	39.9927	.08500	7.9151
22	3.6035	43.3923	.08305	8.2166
23	3.8197	46.9958	.08128	8.5099
24	4.0489	50.8156	.07968	8.7951
25	4.2919	54.8645	.07823	9.0722
26	4.5494	59.1564	.07690	9.3414
27	4.8223	63.7058	.07570	9.6029
28	5.1117	68.5281	.07459	9.8568
29	5.4184	73.6398	.07358	10.1032
30	5.7435	79.0582	.07265	10.3422
31	6.0881	84.8017	.07179	10.5740
32	6.4534	90.8898	.07100	10.7988
33	6.8406	97.3432	.07027	11.0166
34	7.2510	104.1838	.06960	11.2276
35	7.6861	111.4348	.06897	11.4319
40	10.2857	154.7620	.06646	12.3590
45	13.7646	212.7435	.06470	13.1413
50	18.4202	290.3359	.06344	13.7964
55	24.6503	394.1720	.06254	14.3411
60	32.9877	533.1282	.06188	14.7909
65	44.1450	719.0829	.06139	15.1601
70	59.0759	967.9322	.06103	15.4613
75	79.0569	1300.9487	.06077	15.7058
80	105.7960	1746.5999	.06057	15.9033
85	141.5789	2342.9817	.06043	16.0620
90	189.4645	3141.0752	.06032	16.1891
95	253.5463	4209.1042	.06024	16.2905
100	339.3021	5638.3681	.06018	16.3711

COMPOUND INTEREST FACTORS - ANNUAL COMPOUNDING

INTEREST RATE = 7.00 PERCENT

N	SINGLE-PAYMENT COMPOUND-AMOUNT FACTOR (F/P)	UNIFORM-SERIES COMPOUND-AMOUNT FACTOR (F/A)	UNIFORM-SERIES CAPITAL-RECOVERY FACTOR (A/P)	GRADIENT SERIES FACTOR (A/G)
1	1.0700	1.0000	1.07000	.0000
2	1.1449	2.0700	.55309	.4831
3	1.2250	3.2149	.38105	.9549
4	1.3108	4.4399	.29523	1.4155
5	1.4026	5.7507	.24389	1.8650
6	1.5007	7.1533	.20980	2.3032
7	1.6058	8.6540	.18555	2.7304
8	1.7182	10.2598	.16747	3.1465
9	1.8385	11.9780	.15349	3.5517
10	1.9672	13.8164	.14238	3.9461
11	2.1049	15.7836	.13336	4.3296
12	2.2522	17.8885	.12590	4.7025
13	2.4098	20.1406	.11965	5.0648
14	2.5785	22.5505	.11434	5.4167
15	2.7590	25.1290	.10979	5.7583
16	2.9522	27.8881	.10586	6.0897
17	3.1588	30.8402	.10243	6.4110
18	3.3799	33.9990	.09941	6.7225
19	3.6165	37.3790	.09675	7.0242
20	3.8697	40.9955	.09439	7.3163
21	4.1406	44.8652	.09229	7.5990
22	4.4304	49.0057	.09041	7.8725
23	4.7405	53.4361	.08871	8.1369
24	5.0724	58.1767	.08719	8.3923
25	5.4274	63.2490	.08581	8.6391
26	5.8074	68.6765	.08456	8.8773
27	6.2139	74.4838	.08343	9.1072
28	6.6488	80.6977	.08239	9.3289
29	7.1143	87.3465	.08145	9.5427
30	7.6123	94.4608	.08059	9.7487
31	8.1451	102.0730	.07980	9.9471
32	8.7153	110.2182	.07907	10.1381
33	9.3253	118.9334	.07841	10.3219
34	9.9781	128.2588	.07780	10.4987
35	10.6766	138.2369	.07723	10.6687
40	14.9745	199.6351	.07501	11.4233
45	21.0025	285.7493	.07350	12.0360
50	29.4570	406.5289	.07246	12.5287
55	41.3150	575.9286	.07174	12.9215
60	57.9464	813.5204	.07123	13.2321
65	81.2729	1146.7552	.07087	13.4760
70	113.9894	1614.1342	.07062	13.6662
75	159.8760	2269.6574	.07044	13.8136
80	224.2344	3189.0627	.07031	13.9273
85	314.5003	4478.5761	.07022	14.0146
90	441.1030	6287.1854	.07016	14.0812
95	618.6697	8823.8535	.07011	14.1319
100	867.7163	12381.6618	.07008	14.1703

COMPOUND INTEREST FACTORS - ANNUAL COMPOUNDING

INTEREST RATE = 8.00 PERCENT

N	SINGLE-PAYMENT COMPOUND-AMOUNT FACTOR (F/P)	UNIFORM-SERIES COMPOUND-AMOUNT FACTOR (F/A)	UNIFORM-SERIES CAPITAL-RECOVERY FACTOR (A/P)	GRADIENT SERIES FACTOR (A/G)
1	1.0800	1.0000	1.08000	.0000
2	1.1664	2.0800	.56077	.4808
3	1.2597	3.2464	.38803	.9487
4	1.3605	4.5061	.30192	1.4040
5	1.4693	5.8666	.25046	1.8465
6	1.5869	7.3359	.21632	2.2763
7	1.7138	8.9228	.19207	2.6937
8	1.8509	10.6366	.17401	3.0985
9	1.9990	12.4876	.16008	3.4910
10	2.1589	14.4866	.14903	3.8713
11	2.3316	16.6455	.14008	4.2395
12	2.5182	18.9771	.13270	4.5957
13	2.7196	21.4953	.12652	4.9402
14	2.9372	24.2149	.12130	5.2731
15	3.1722	27.1521	.11683	5.5945
16	3.4259	30.3243	.11298	5.9046
17	3.7000	33.7502	.10963	6.2037
18	3.9960	37.4502	.10670	6.4920
19	4.3157	41.4463	.10413	6.7697
20	4.6610	45.7620	.10185	7.0369
21	5.0338	50.4229	.09983	7.2940
22	5.4365	55.4568	.09803	7.5412
23	5.8715	60.8933	.09642	7.7786
24	6.3412	66.7648	.09498	8.0066
25	6.8485	73.1059	.09368	8.2254
26	7.3964	79.9544	.09251	8.4352
27	7.9881	87.3508	.09145	8.6363
28	8.6271	95.3388	.09049	8.8289
29	9.3173	103.9659	.08962	9.0133
30	10.0627	113.2832	.08883	9.1897
31	10.8677	123.3459	.08811	9.3584
32	11.7371	134.2135	.08745	9.5197
33	12.6760	145.9506	.08685	9.6737
34	13.6901	158.6267	.08630	9.8208
35	14.7853	172.3168	.08580	9.9611
40	21.7245	259.0565	.08386	10.5699
45	31.9204	386.5056	.08259	11.0447
50	46.9016	573.7702	.08174	11.4107
55	68.9139	848.9232	.08118	11.6902
60	101.2571	1253.2133	.08080	11.9015
65	148.7798	1847.2481	.08054	12.0602
70	218.6064	2720.0801	.08037	12.1783
75	321.2045	4002.5566	.08025	12.2658
80	471.9548	5886.9354	.08017	12.3301
85	693.4565	8655.7061	.08012	12.3772
90	1018.9151	12723.9386	.08008	12.4116
95	1497.1205	18701.5069	.08005	12.4365
100	2199.7613	27484.5157	.08004	12.4545

COMPOUND INTEREST FACTORS - ANNUAL COMPOUNDING

INTEREST RATE = 9.00 PERCENT

N	SINGLE-PAYMENT COMPOUND-AMOUNT FACTOR (F/P)	UNIFORM-SERIES COMPOUND-AMOUNT FACTOR (F/A)	UNIFORM-SERIES CAPITAL-RECOVERY FACTOR (A/P)	GRADIENT SERIES FACTOR (A/G)
1	1.0900	1.0000	1.09000	.0000
2	1.1881	2.0900	.56847	.4785
3	1.2950	3.2781	.39505	.9426
4	1.4116	4.5731	.30867	1.3925
5	1.5386	5.9847	.25709	1.8282
6	1.6771	7.5233	.22292	2.2498
7	1.8280	9.2004	.19869	2.6574
8	1.9926	11.0285	.18067	3.0512
9	2.1719	13.0210	.16680	3.4312
10	2.3674	15.1929	.15582	3.7978
11	2.5804	17.5603	.14695	4.1510
12	2.8127	20.1407	.13965	4.4910
13	3.0658	22.9534	.13357	4.8182
14	3.3417	26.0192	.12843	5.1326
15	3.6425	29.3609	.12406	5.4346
16	3.9703	33.0034	.12030	5.7245
17	4.3276	36.9737	.11705	6.0024
18	4.7171	41.3013	.11421	6.2687
19	5.1417	46.0185	.11173	6.5236
20	5.6044	51.1601	.10955	6.7674
21	6.1088	56.7645	.10762	7.0006
22	6.6586	62.8733	.10590	7.2232
23	7.2579	69.5319	.10438	7.4357
24	7.9111	76.7898	.10302	7.6384
25	8.6231	84.7009	.10181	7.8316
26	9.3992	93.3240	.10072	8.0156
27	10.2451	102.7231	.09973	8.1906
28	11.1671	112.9682	.09885	8.3571
29	12.1722	124.1354	.09806	8.5154
30	13.2677	136.3075	.09734	8.6657
31	14.4618	149.5752	.09669	8.8083
32	15.7633	164.0370	.09610	8.9436
33	17.1820	179.8003	.09556	9.0718
34	18.7284	196.9823	.09508	9.1933
35	20.4140	215.7108	.09464	9.3083
40	31.4094	337.8824	.09296	9.7957
45	48.3273	525.8587	.09190	10.1603
50	74.3575	815.0836	.09123	10.4295
55	114.4083	1260.0918	.09079	10.6261
60	176.0313	1944.7921	.09051	10.7683
65	270.8460	2998.2885	.09033	10.8702
70	416.7301	4619.2232	.09022	10.9427
75	641.1909	7113.2321	.09014	10.9940
80	986.5517	10950.5741	.09009	11.0299
85	1517.9320	16854.8003	.09006	11.0551
90	2335.5266	25939.1842	.09004	11.0726
95	3593.4971	39916.6350	.09003	11.0847
100	5529.0408	61422.6755	.09002	11.0930

COMPOUND INTEREST FACTORS - ANNUAL COMPOUNDING

INTEREST RATE = 10.00 PERCENT

N	SINGLE-PAYMENT COMPOUND-AMOUNT FACTOR (F/P)	UNIFORM-SERIES COMPOUND-AMOUNT FACTOR (F/A)	UNIFORM-SERIES CAPITAL-RECOVERY FACTOR (A/P)	GRADIENT SERIES FACTOR (A/G)
1	1.1000	1.0000	1.10000	.0000
2	1.2100	2.1000	.57619	.4762
3	1.3310	3.3100	.40211	.9366
4	1.4641	4.6410	.31547	1.3812
5	1.6105	6.1051	.26380	1.8101
6	1.7716	7.7156	.22961	2.2236
7	1.9487	9.4872	.20541	2.6216
8	2.1436	11.4359	.18744	3.0045
9	2.3579	13.5795	.17364	3.3724
10	2.5937	15.9374	.16275	3.7255
11	2.8531	18.5312	.15396	4.0641
12	3.1384	21.3843	.14676	4.3884
13	3.4523	24.5227	.14078	4.6988
14	3.7975	27.9750	.13575	4.9955
15	4.1772	31.7725	.13147	5.2789
16	4.5950	35.9497	.12782	5.5493
17	5.0545	40.5447	.12466	5.8071
18	5.5599	45.5992	.12193	6.0526
19	6.1159	51.1591	.11955	6.2861
20	6.7275	57.2750	.11746	6.5081
21	7.4002	64.0025	.11562	6.7189
22	8.1403	71.4027	.11401	6.9189
23	8.9543	79.5430	.11257	7.1085
24	9.8497	88.4973	.11130	7.2881
25	10.8347	98.3471	.11017	7.4580
26	11.9182	109.1818	.10916	7.6186
27	13.1100	121.0999	.10826	7.7704
28	14.4210	134.2099	.10745	7.9137
29	15.8631	148.6309	.10673	8.0489
30	17.4494	164.4940	.10608	8.1762
31	19.1943	181.9434	.10550	8.2962
32	21.1138	201.1378	.10497	8.4091
33	23.2252	222.2515	.10450	8.5152
34	25.5477	245.4767	.10407	8.6149
35	28.1024	271.0244	.10369	8.7086
40	45.2593	442.5926	.10226	9.0962
45	72.8905	718.9048	.10139	9.3740
50	117.3909	1163.9085	.10086	9.5704
55	189.0591	1880.5914	.10053	9.7075
60	304.4816	3034.8164	.10033	9.8023
65	490.3707	4893.7073	.10020	9.8672
70	789.7470	7887.4696	.10013	9.9113
75	1271.8954	12708.9537	.10008	9.9410
80	2048.4002	20474.0021	.10005	9.9609
85	3298.9690	32979.6903	.10003	9.9742
90	5313.0226	53120.2261	.10002	9.9831
95	8556.6760	85556.7605	.10001	9.9889
100	13780.6123	137796.1234	.10001	9.9927

COMPOUND INTEREST FACTORS - ANNUAL COMPOUNDING

INTEREST RATE = 12.00 PERCENT

N	SINGLE-PAYMENT COMPOUND-AMOUNT FACTOR (F/P)	UNIFORM-SERIES COMPOUND-AMOUNT FACTOR (F/A)	UNIFORM-SERIES CAPITAL-RECOVERY FACTOR (A/P)	GRADIENT SERIES FACTOR (A/G)
1	1.1200	1.0000	1.12000	.0000
2	1.2544	2.1200	.59170	.4717
3	1.4049	3.3744	.41635	.9246
4	1.5735	4.7793	.32923	1.3589
5	1.7623	6.3528	.27741	1.7746
6	1.9738	8.1152	.24323	2.1720
7	2.2107	10.0890	.21912	2.5515
8	2.4760	12.2997	.20130	2.9131
9	2.7731	14.7757	.18768	3.2574
10	3.1058	17.5487	.17698	3.5847
11	3.4785	20.6546	.16842	3.8953
12	3.8960	24.1331	.16144	4.1897
13	4.3635	28.0291	.15568	4.4683
14	4.8871	32.3926	.15087	4.7317
15	5.4736	37.2797	.14682	4.9803
16	6.1304	42.7533	.14339	5.2147
17	6.8660	48.8837	.14046	5.4353
18	7.6900	55.7497	.13794	5.6427
19	8.6128	63.4397	.13576	5.8375
20	9.6463	72.0524	.13388	6.0202
21	10.8038	81.6987	.13224	6.1913
22	12.1003	92.5026	.13081	6.3514
23	13.5523	104.6029	.12956	6.5010
24	15.1786	118.1552	.12846	6.6406
25	17.0001	133.3339	.12750	6.7708
26	19.0401	150.3339	.12665	6.8921
27	21.3249	169.3740	.12590	7.0049
28	23.8839	190.6989	.12524	7.1098
29	26.7499	214.5828	.12466	7.2071
30	29.9599	241.3327	.12414	7.2974
31	33.5551	271.2926	.12369	7.3811
32	37.5817	304.8477	.12328	7.4586
33	42.0915	342.4294	.12292	7.5302
34	47.1425	384.5210	.12260	7.5965
35	52.7996	431.6635	.12232	7.6577
40	93.0510	767.0914	.12130	7.8988
45	163.9876	1358.2300	.12074	8.0572
50	289.0022	2400.0182	.12042	8.1597

COMPOUND INTEREST FACTORS - ANNUAL COMPOUNDING

INTEREST RATE = 15.00 PERCENT

N	SINGLE-PAYMENT COMPOUND-AMOUNT FACTOR (F/P)	UNIFORM-SERIES COMPOUND-AMOUNT FACTOR (F/A)	UNIFORM-SERIES CAPITAL-RECOVERY FACTOR (A/P)	GRADIENT SERIES FACTOR (A/G)
1	1.1500	1.0000	1.15000	.0000
2	1.3225	2.1500	.61512	.4651
3	1.5209	3.4725	.43798	.9071
4	1.7490	4.9934	.35027	1.3263
5	2.0114	6.7424	.29832	1.7228
6	2.3131	8.7537	.26424	2.0972
7	2.6600	11.0668	.24036	2.4498
8	3.0590	13.7268	.22285	2.7813
9	3.5179	16.7858	.20957	3.0922
10	4.0456	20.3037	.19925	3.3832
11	4.6524	24.3493	.19107	3.6549
12	5.3503	29.0017	.18448	3.9082
13	6.1528	34.3519	.17911	4.1438
14	7.0757	40.5047	.17469	4.3624
15	8.1371	47.5804	.17102	4.5650
16	9.3576	55.7175	.16795	4.7522
17	10.7613	65.0751	.16537	4.9251
18	12.3755	75.8364	.16319	5.0843
19	14.2318	88.2118	.16134	5.2307
20	16.3665	102.4436	.15976	5.3651
21	18.8215	118.8101	.15842	5.4883
22	21.6447	137.6316	.15727	5.6010
23	24.8915	159.2764	.15628	5.7040
24	28.6252	184.1678	.15543	5.7979
25	32.9190	212.7930	.15470	5.8834
26	37.8568	245.7120	.15407	5.9612
27	43.5353	283.5688	.15353	6.0319
28	50.0656	327.1041	.15306	6.0960
29	57.5755	377.1697	.15265	6.1541
30	66.2118	434.7451	.15230	6.2066
31	76.1435	500.9569	.15200	6.2541
32	87.5651	577.1005	.15173	6.2970
33	100.6998	664.6655	.15150	6.3357
34	115.8048	765.3654	.15131	6.3705
35	133.1755	881.1702	.15113	6.4019
40	267.8635	1779.0903	.15056	6.5168
45	538.7693	3585.1285	.15028	6.5830
50	1083.6574	7217.7163	.15014	6.6205

COMPOUND INTEREST FACTORS - ANNUAL COMPOUNDING

INTEREST RATE = 20.00 PERCENT

N	SINGLE-PAYMENT COMPOUND-AMOUNT FACTOR (F/P)	UNIFORM-SERIES COMPOUND-AMOUNT FACTOR (F/A)	UNIFORM-SERIES CAPITAL-RECOVERY FACTOR (A/P)	GRADIENT SERIES FACTOR (A/G)
1	1.2000	1.0000	1.20000	.0000
2	1.4400	2.2000	.65455	.4545
3	1.7280	3.6400	.47473	.8791
4	2.0736	5.3680	.38629	1.2742
5	2.4883	7.4416	.33438	1.6405
6	2.9860	9.9299	.30071	1.9788
7	3.5832	12.9159	.27742	2.2902
8	4.2998	16.4991	.26061	2.5756
9	5.1598	20.7989	.24808	2.8364
10	6.1917	25.9587	.23852	3.0739
11	7.4301	32.1504	.23110	3.2893
12	8.9161	39.5805	.22526	3.4841
13	10.6993	48.4966	.22062	3.6597
14	12.8392	59.1959	.21689	3.8175
15	15.4070	72.0351	.21388	3.9588
16	18.4884	87.4421	.21144	4.0851
17	22.1861	105.9306	.20944	4.1976
18	26.6233	128.1167	.20781	4.2975
19	31.9480	154.7400	.20646	4.3861
20	38.3376	186.6880	.20536	4.4643
21	46.0051	225.0256	.20444	4.5334
22	55.2061	271.0307	.20369	4.5941
23	66.2474	326.2369	.20307	4.6475
24	79.4968	392.4842	.20255	4.6943
25	95.3962	471.9811	.20212	4.7352
26	114.4755	567.3773	.20176	4.7709
27	137.3706	681.8528	.20147	4.8020
28	164.8447	819.2233	.20122	4.8291
29	197.8136	984.0680	.20102	4.8527
30	237.3763	1181.8816	.20085	4.8731
31	284.8516	1419.2579	.20070	4.8908
32	341.8219	1704.1095	.20059	4.9061
33	410.1863	2045.9314	.20049	4.9194
34	492.2235	2456.1176	.20041	4.9308
35	590.6682	2948.3411	.20034	4.9406
40	1469.7716	7343.8578	.20014	4.9728
45	3657.2620	18281.3099	.20005	4.9877
50	9100.4382	45497.1908	.20002	4.9945

COMPOUND INTEREST FACTORS - ANNUAL COMPOUNDING

INTEREST RATE = 25.00 PERCENT

N	SINGLE-PAYMENT COMPOUND-AMOUNT FACTOR (F/P)	UNIFORM-SERIES COMPOUND-AMOUNT FACTOR (F/A)	UNIFORM-SERIES CAPITAL-RECOVERY FACTOR (A/P)	GRADIENT SERIES FACTOR (A/G)
1	1.2500	1.0000	1.25000	.0000
2	1.5625	2.2500	.69444	.4444
3	1.9531	3.8125	.51230	.8525
4	2.4414	5.7656	.42344	1.2249
5	3.0518	8.2070	.37185	1.5631
6	3.8147	11.2588	.33882	1.8683
7	4.7684	15.0735	.31634	2.1424
8	5.9605	19.8419	.30040	2.3872
9	7.4506	25.8023	.28876	2.6048
10	9.3132	33.2529	.28007	2.7971
11	11.6415	42.5661	.27349	2.9663
12	14.5519	54.2077	.26845	3.1145
13	18.1899	68.7596	.26454	3.2437
14	22.7374	86.9495	.26150	3.3559
15	28.4217	109.6868	.25912	3.4530
16	35.5271	138.1085	.25724	3.5366
17	44.4089	173.6357	.25576	3.6084
18	55.5112	218.0446	.25459	3.6698
19	69.3889	273.5558	.25366	3.7222
20	86.7362	342.9447	.25292	3.7667
21	108.4202	429.6809	.25233	3.8045
22	135.5253	538.1011	.25186	3.8365
23	169.4066	673.6264	.25148	3.8634
24	211.7582	843.0329	.25119	3.8861
25	264.6978	1054.7912	.25095	3.9052
26	330.8722	1319.4890	.25076	3.9212
27	413.5903	1650.3612	.25061	3.9346
28	516.9879	2063.9515	.25048	3.9457
29	646.2349	2580.9394	.25039	3.9551
30	807.7936	3227.1743	.25031	3.9628
31	1009.7420	4034.9678	.25025	3.9693
32	1262.1774	5044.7098	.25020	3.9746
33	1577.7218	6306.8872	.25016	3.9791
34	1972.1523	7884.6091	.25013	3.9828
35	2465.1903	9856.7613	.25010	3.9858

COMPOUND INTEREST FACTORS - ANNUAL COMPOUNDING

INTEREST RATE = 30.00 PERCENT

N	SINGLE-PAYMENT COMPOUND-AMOUNT FACTOR (F/P)	UNIFORM-SERIES COMPOUND-AMOUNT FACTOR (F/A)	UNIFORM-SERIES CAPITAL-RECOVERY FACTOR (A/P)	GRADIENT SERIES FACTOR (A/G)
1	1.3000	1.0000	1.30000	.0000
2	1.6900	2.3000	.73478	.4348
3	2.1970	3.9900	.55063	.8271
4	2.8561	6.1870	.46163	1.1783
5	3.7129	9.0431	.41058	1.4903
6	4.8268	12.7560	.37839	1.7654
7	6.2749	17.5828	.35687	2.0063
8	8.1573	23.8577	.34192	2.2156
9	10.6045	32.0150	.33124	2.3963
10	13.7858	42.6195	.32346	2.5512
11	17.9216	56.4053	.31773	2.6833
12	23.2981	74.3270	.31345	2.7952
13	30.2875	97.6250	.31024	2.8895
14	39.3738	127.9125	.30782	2.9685
15	51.1859	167.2863	.30598	3.0344
16	66.5417	218.4722	.30458	3.0892
17	86.5042	285.0139	.30351	3.1345
18	112.4554	371.5180	.30269	3.1718
19	146.1920	483.9734	.30207	3.2025
20	190.0496	630.1655	.30159	3.2275
21	247.0645	820.2151	.30122	3.2480
22	321.1839	1067.2796	.30094	3.2646
23	417.5391	1388.4635	.30072	3.2781
24	542.8008	1806.0026	.30055	3.2890
25	705.6410	2348.8033	.30043	3.2979
26	917.3333	3054.4443	.30033	3.3050
27	1192.5333	3971.7776	.30025	3.3107
28	1550.2933	5164.3109	.30019	3.3153
29	2015.3813	6714.6042	.30015	3.3189
30	2619.9956	8729.9855	.30011	3.3219
31	3405.9943	11349.9811	.30009	3.3242
32	4427.7926	14755.9755	.30007	3.3261
33	5756.1304	19183.7681	.30005	3.3276
34	7482.9696	24939.8985	.30004	3.3288
35	9727.8604	32422.8681	.30003	3.3297

COMPOUND INTEREST FACTORS - ANNUAL COMPOUNDING

INTEREST RATE = 40.00 PERCENT

N	SINGLE-PAYMENT COMPOUND-AMOUNT FACTOR (F/P)	UNIFORM-SERIES COMPOUND-AMOUNT FACTOR (F/A)	UNIFORM-SERIES CAPITAL-RECOVERY FACTOR (A/P)	GRADIENT SERIES FACTOR (A/G)
1	1.4000	1.0000	1.40000	.0000
2	1.9600	2.4000	.81667	.4167
3	2.7440	4.3600	.62936	.7798
4	3.8416	7.1040	.54077	1.0923
5	5.3782	10.9456	.49136	1.3580
6	7.5295	16.3238	.46126	1.5811
7	10.5414	23.8534	.44192	1.7664
8	14.7579	34.3947	.42907	1.9185
9	20.6610	49.1526	.42034	2.0422
10	28.9255	69.8137	.41432	2.1419
11	40.4957	98.7391	.41013	2.2215
12	56.6939	139.2348	.40718	2.2845
13	79.3715	195.9287	.40510	2.3341
14	111.1201	275.3002	.40363	2.3729
15	155.5681	386.4202	.40259	2.4030
16	217.7953	541.9883	.40185	2.4262
17	304.9135	759.7837	.40132	2.4441
18	426.8789	1064.6971	.40094	2.4577
19	597.6304	1491.5760	.40067	2.4682
20	836.6826	2089.2064	.40048	2.4761
21	1171.3556	2925.8889	.40034	2.4821
22	1639.8978	4097.2445	.40024	2.4866
23	2295.8569	5737.1423	.40017	2.4900
24	3214.1997	8032.9993	.40012	2.4925
25	4499.8796	11247.1990	.40009	2.4944
26	6299.8314	15747.0785	.40006	2.4959
27	8819.7640	22046.9099	.40005	2.4969
28	12347.6696	30866.6739	.40003	2.4977
29	17286.7374	43214.3435	.40002	2.4983
30	24201.4324	60501.0809	.40002	2.4988

COMPOUND INTEREST FACTORS - ANNUAL COMPOUNDING

INTEREST RATE = 50.00 PERCENT

N	SINGLE-PAYMENT COMPOUND-AMOUNT FACTOR (F/P)	UNIFORM-SERIES COMPOUND-AMOUNT FACTOR (F/A)	UNIFORM-SERIES CAPITAL-RECOVERY FACTOR (A/P)	GRADIENT SERIES FACTOR (A/G)
1	1.5000	1.0000	1.50000	.0000
2	2.2500	2.5000	.90000	.4000
3	3.3750	4.7500	.71053	.7368
4	5.0625	8.1250	.62308	1.0154
5	7.5938	13.1875	.57583	1.2417
6	11.3906	20.7813	.54812	1.4226
7	17.0859	32.1719	.53108	1.5648
8	25.6289	49.2578	.52030	1.6752
9	38.4434	74.8867	.51335	1.7596
10	57.6650	113.3301	.50882	1.8235
11	86.4976	170.9951	.50585	1.8713
12	129.7463	257.4927	.50388	1.9068
13	194.6195	387.2390	.50258	1.9329
14	291.9293	581.8585	.50172	1.9519
15	437.8939	873.7878	.50114	1.9657
16	656.8408	1311.6817	.50076	1.9756
17	985.2613	1968.5225	.50051	1.9827
18	1477.8919	2953.7838	.50034	1.9878
19	2216.8378	4431.6756	.50023	1.9914
20	3325.2567	6648.5135	.50015	1.9940
21	4987.8851	9973.7702	.50010	1.9958
22	7481.8276	14961.6553	.50007	1.9971
23	11222.7415	22443.4829	.50004	1.9980
24	16834.1122	33666.2244	.50003	1.9986
25	25251.1683	50500.3366	.50002	1.9990

Appendix B

Nominal versus Effective Interest Rates

EFFECTIVE INTEREST RATES

NOMINAL INTEREST RATE	FREQUENCY OF COMPOUNDING					
	SEMIANNUALLY	QUARTERLY	MONTHLY	WEEKLY	DAILY	CONTINUOUSLY
.01	.010025	.010038	.010046	.010049	.010050	.010050
.02	.020100	.020151	.020184	.020197	.020201	.020201
.03	.030225	.030339	.030416	.030446	.030453	.030455
.04	.040400	.040604	.040742	.040795	.040808	.040811
.05	.050625	.050945	.051162	.051246	.051267	.051271
.06	.060900	.061364	.061678	.061800	.061831	.061837
.07	.071225	.071859	.072290	.072458	.072501	.072508
.08	.081600	.082432	.083000	.083220	.083278	.083287
.09	.092025	.093083	.093807	.094089	.094162	.094174
.10	.102500	.103813	.104713	.105065	.105156	.105171
.11	.113025	.114621	.115719	.116148	.116260	.116278
.12	.123600	.125509	.126825	.127341	.127475	.127497
.13	.134225	.136476	.138032	.138644	.138802	.138828
.14	.144900	.147523	.149342	.150057	.150243	.150274
.15	.155625	.158650	.160755	.161583	.161798	.161834
.16	.166400	.169859	.172271	.173223	.173470	.173511
.17	.177225	.181148	.183892	.184976	.185258	.185305
.18	.188100	.192519	.195618	.196845	.197164	.197217
.19	.199025	.203971	.207451	.208831	.209190	.209250
.20	.210000	.215506	.219391	.220934	.221336	.221403
.21	.221025	.227124	.231439	.233156	.233604	.233678
.22	.232100	.238825	.243597	.245499	.245994	.246077
.23	.243225	.250609	.255864	.257962	.258509	.258600
.24	.254400	.262477	.268242	.270547	.271149	.271249
.25	.265625	.274429	.280732	.283256	.283916	.284025
.26	.276900	.286466	.293334	.296090	.296810	.296930
.27	.288225	.298588	.306050	.309050	.309834	.309964
.28	.299600	.310796	.318881	.322136	.322988	.323130
.29	.311025	.323089	.331826	.335351	.336274	.336427
.30	.322500	.335469	.344889	.348696	.349692	.349859
.31	.334025	.347936	.358069	.362171	.363246	.363425
.32	.345600	.360489	.371367	.375778	.376935	.377128
.33	.357225	.373130	.384784	.389519	.390761	.390968
.34	.368900	.385859	.398321	.403394	.404725	.404948
.35	.380625	.398676	.411980	.417404	.418830	.419068
.36	.392400	.411582	.425761	.431553	.433075	.433329
.37	.404225	.424577	.439665	.445839	.447463	.447735
.38	.416100	.437661	.453693	.460265	.461996	.462285
.39	.428025	.450835	.467847	.474833	.476673	.476981
.40	.440000	.464100	.482126	.489543	.491498	.491825
.41	.452025	.477455	.496533	.504397	.506471	.506818
.42	.464100	.490902	.511069	.519396	.521594	.521962
.43	.476225	.504440	.525733	.534542	.536869	.537258
.44	.488400	.518070	.540528	.549836	.552296	.552707
.45	.500625	.531793	.555454	.565279	.567878	.568312
.46	.512900	.545608	.570513	.580873	.583615	.584074
.47	.525225	.559517	.585705	.596620	.599511	.599994
.48	.537600	.573519	.601032	.612520	.615565	.616074
.49	.550025	.587616	.616495	.628576	.631780	.632316
.50	.562500	.601807	.632094	.644788	.648157	.648721

Appendix C

Compound Interest Factors—Continuous Compounding

COMPOUND INTEREST FACTORS - CONTINUOUS COMPOUNDING

NOMINAL INTEREST RATE = 0.25 PERCENT

N	SINGLE-PAYMENT COMPOUND-AMOUNT FACTOR (F/P)	UNIFORM-SERIES COMPOUND-AMOUNT FACTOR (F/A)	UNIFORM-SERIES CAPITAL-RECOVERY FACTOR (A/P)	GRADIENT SERIES FACTOR (A/G)
1	1.0025	1.0000	1.00250	.0000
2	1.0050	2.0025	.50188	.4994
3	1.0075	3.0075	.33500	.9983
4	1.0101	4.0150	.25157	1.4969
5	1.0126	5.0251	.20150	1.9950
6	1.0151	6.0377	.16813	2.4927
7	1.0177	7.0528	.14429	2.9900
8	1.0202	8.0704	.12641	3.4869
9	1.0228	9.0906	.11251	3.9833
10	1.0253	10.1134	.10138	4.4794
11	1.0279	11.1387	.09228	4.9750
12	1.0305	12.1666	.08470	5.4702
13	1.0330	13.1970	.07828	5.9650
14	1.0356	14.2301	.07278	6.4594
15	1.0382	15.2657	.06801	6.9533
16	1.0408	16.3039	.06384	7.4469
17	1.0434	17.3447	.06016	7.9400
18	1.0460	18.3881	.05689	8.4327
19	1.0486	19.4342	.05396	8.9250
20	1.0513	20.4828	.05132	9.4169
21	1.0539	21.5341	.04894	9.9083
22	1.0565	22.5880	.04677	10.3994
23	1.0592	23.6445	.04480	10.8900
24	1.0618	24.7037	.04298	11.3802
25	1.0645	25.7655	.04131	11.8700
26	1.0672	26.8300	.03977	12.3594
27	1.0698	27.8972	.03835	12.8483
28	1.0725	28.9670	.03703	13.3369
29	1.0752	30.0395	.03579	13.8250
30	1.0779	31.1147	.03464	14.3127
31	1.0806	32.1926	.03357	14.8000
32	1.0833	33.2732	.03256	15.2869
33	1.0860	34.3565	.03161	15.7734
34	1.0887	35.4425	.03072	16.2594
35	1.0914	36.5312	.02988	16.7450
40	1.1052	42.0158	.02630	19.1669
45	1.1191	47.5694	.02353	21.5784
50	1.1331	53.1928	.02130	23.9795
55	1.1474	58.8870	.01948	26.3702
60	1.1618	64.6528	.01797	28.7505
65	1.1764	70.4911	.01669	31.1204
70	1.1912	76.4029	.01559	33.4799
75	1.2062	82.3890	.01464	35.8290
80	1.2214	88.4504	.01381	38.1678
85	1.2368	94.5881	.01308	40.4961
90	1.2523	100.8030	.01242	42.8141
95	1.2681	107.0960	.01184	45.1218
100	1.2840	113.4682	.01132	47.4190

COMPOUND INTEREST FACTORS - CONTINUOUS COMPOUNDING

NOMINAL INTEREST RATE = 0.50 PERCENT

N	SINGLE-PAYMENT COMPOUND-AMOUNT FACTOR (F/P)	UNIFORM-SERIES COMPOUND-AMOUNT FACTOR (F/A)	UNIFORM-SERIES CAPITAL-RECOVERY FACTOR (A/P)	GRADIENT SERIES FACTOR (A/G)
1	1.0050	1.0000	1.00501	.0000
2	1.0101	2.0050	.50376	.4988
3	1.0151	3.0151	.33668	.9967
4	1.0202	4.0302	.25314	1.4938
5	1.0253	5.0504	.20302	1.9900
6	1.0305	6.0757	.16960	2.4854
7	1.0356	7.1061	.14574	2.9800
8	1.0408	8.1418	.12784	3.4738
9	1.0460	9.1826	.11391	3.9667
10	1.0513	10.2286	.10278	4.4588
11	1.0565	11.2799	.09367	4.9500
12	1.0618	12.3364	.08607	5.4404
13	1.0672	13.3983	.07965	5.9300
14	1.0725	14.4654	.07414	6.4188
15	1.0779	15.5379	.06937	6.9067
16	1.0833	16.6158	.06520	7.3938
17	1.0887	17.6991	.06151	7.8800
18	1.0942	18.7878	.05824	8.3654
19	1.0997	19.8820	.05531	8.8500
20	1.1052	20.9816	.05267	9.3338
21	1.1107	22.0868	.05029	9.8167
22	1.1163	23.1975	.04812	10.2988
23	1.1219	24.3138	.04614	10.7800
24	1.1275	25.4357	.04433	11.2605
25	1.1331	26.5632	.04266	11.7401
26	1.1388	27.6963	.04112	12.2188
27	1.1445	28.8351	.03969	12.6968
28	1.1503	29.9797	.03837	13.1739
29	1.1560	31.1300	.03714	13.6501
30	1.1618	32.2860	.03599	14.1256
31	1.1677	33.4478	.03491	14.6002
32	1.1735	34.6155	.03390	15.0739
33	1.1794	35.7890	.03295	15.5469
34	1.1853	36.9684	.03206	16.0190
35	1.1912	38.1537	.03122	16.4903
40	1.2214	44.1699	.02765	18.8342
45	1.2523	50.3385	.02488	21.1574
50	1.2840	56.6632	.02266	23.4598
55	1.3165	63.1480	.02085	25.7416
60	1.3499	69.7970	.01934	28.0027
65	1.3840	76.6143	.01806	30.2431
70	1.4191	83.6042	.01697	32.4629
75	1.4550	90.7710	.01603	34.6621
80	1.4918	98.1192	.01520	36.8408
85	1.5296	105.6535	.01448	38.9990
90	1.5683	113.3785	.01383	41.1368
95	1.6080	121.2991	.01326	43.2541
100	1.6487	129.4202	.01274	45.3510

COMPOUND INTEREST FACTORS - CONTINUOUS COMPOUNDING

NOMINAL INTEREST RATE = 0.75 PERCENT

N	SINGLE-PAYMENT COMPOUND-AMOUNT FACTOR (F/P)	UNIFORM-SERIES COMPOUND-AMOUNT FACTOR (F/A)	UNIFORM-SERIES CAPITAL-RECOVERY FACTOR (A/P)	GRADIENT SERIES FACTOR (A/G)
1	1.0075	1.0000	1.00753	.0000
2	1.0151	2.0075	.50565	.4981
3	1.0228	3.0226	.33836	.9950
4	1.0305	4.0454	.25472	1.4906
5	1.0382	5.0759	.20454	1.9850
6	1.0460	6.1141	.17109	2.4781
7	1.0539	7.1601	.14719	2.9700
8	1.0618	8.2140	.12927	3.4606
9	1.0698	9.2758	.11534	3.9500
10	1.0779	10.3457	.10419	4.4381
11	1.0860	11.4235	.09507	4.9250
12	1.0942	12.5095	.08747	5.4106
13	1.1024	13.6037	.08104	5.8950
14	1.1107	14.7061	.07553	6.3781
15	1.1191	15.8168	.07075	6.8600
16	1.1275	16.9359	.06657	7.3407
17	1.1360	18.0634	.06289	7.8200
18	1.1445	19.1994	.05961	8.2982
19	1.1532	20.3439	.05668	8.7751
20	1.1618	21.4971	.05405	9.2507
21	1.1706	22.6589	.05166	9.7251
22	1.1794	23.8295	.04949	10.1983
23	1.1883	25.0089	.04751	10.6702
24	1.1972	26.1972	.04570	11.1408
25	1.2062	27.3944	.04403	11.6102
26	1.2153	28.6006	.04249	12.0784
27	1.2245	29.8159	.04107	12.5453
28	1.2337	31.0404	.03974	13.0110
29	1.2430	32.2741	.03851	13.4754
30	1.2523	33.5170	.03736	13.9386
31	1.2618	34.7693	.03629	14.4005
32	1.2712	36.0311	.03528	14.8612
33	1.2808	37.3023	.03434	15.3207
34	1.2905	38.5832	.03345	15.7789
35	1.3002	39.8736	.03261	16.2359
40	1.3499	46.4731	.02905	18.5021
45	1.4014	53.3248	.02628	20.7374
50	1.4550	60.4383	.02407	22.9418
55	1.5106	67.8236	.02227	25.1153
60	1.5683	75.4912	.02077	27.2582
65	1.6282	83.4517	.01951	29.3704
70	1.6905	91.7164	.01843	31.4521
75	1.7551	100.2969	.01750	33.5034
80	1.8221	109.2053	.01669	35.5244
85	1.8917	118.4541	.01597	37.5153
90	1.9640	128.0563	.01534	39.4762
95	2.0391	138.0255	.01477	41.4072
100	2.1170	148.3755	.01427	43.3084

COMPOUND INTEREST FACTORS - CONTINUOUS COMPOUNDING

NOMINAL INTEREST RATE = 1.00 PERCENT

N	SINGLE-PAYMENT COMPOUND-AMOUNT FACTOR (F/P)	UNIFORM-SERIES COMPOUND-AMOUNT FACTOR (F/A)	UNIFORM-SERIES CAPITAL-RECOVERY FACTOR (A/P)	GRADIENT SERIES FACTOR (A/G)
1	1.0101	1.0000	1.01005	.0000
2	1.0202	2.0101	.50755	.4975
3	1.0305	3.0303	.34006	.9933
4	1.0408	4.0607	.25631	1.4875
5	1.0513	5.1015	.20607	1.9800
6	1.0618	6.1528	.17258	2.4708
7	1.0725	7.2146	.14866	2.9600
8	1.0833	8.2871	.13072	3.4475
9	1.0942	9.3704	.11677	3.9333
10	1.1052	10.4646	.10561	4.4175
11	1.1163	11.5698	.09648	4.9000
12	1.1275	12.6860	.08888	5.3809
13	1.1388	13.8135	.08244	5.8600
14	1.1503	14.9524	.07693	6.3376
15	1.1618	16.1026	.07215	6.8134
16	1.1735	17.2645	.06797	7.2876
17	1.1853	18.4380	.06429	7.7601
18	1.1972	19.6233	.06101	8.2310
19	1.2092	20.8205	.05808	8.7002
20	1.2214	22.0298	.05544	9.1677
21	1.2337	23.2512	.05306	9.6336
22	1.2461	24.4848	.05089	10.0978
23	1.2586	25.7309	.04891	10.5604
24	1.2712	26.9895	.04710	11.0213
25	1.2840	28.2608	.04543	11.4805
26	1.2969	29.5448	.04390	11.9381
27	1.3100	30.8417	.04247	12.3941
28	1.3231	32.1517	.04115	12.8484
29	1.3364	33.4748	.03992	13.3010
30	1.3499	34.8112	.03878	13.7520
31	1.3634	36.1611	.03770	14.2013
32	1.3771	37.5245	.03670	14.6490
33	1.3910	38.9017	.03576	15.0950
34	1.4049	40.2926	.03487	15.5394
35	1.4191	41.6976	.03403	15.9821
40	1.4918	48.9370	.03048	18.1710
45	1.5683	56.5475	.02773	20.3190
50	1.6487	64.5483	.02554	22.4261
55	1.7333	72.9593	.02376	24.4926
60	1.8221	81.8015	.02227	26.5187
65	1.9155	91.0971	.02103	28.5045
70	2.0138	100.8692	.01996	30.4505
75	2.1170	111.1424	.01905	32.3567
80	2.2255	121.9423	.01825	34.2235
85	2.3396	133.2960	.01755	36.0513
90	2.4596	145.2317	.01694	37.8402
95	2.5857	157.7794	.01639	39.5907
100	2.7183	170.9705	.01590	41.3032

COMPOUND INTEREST FACTORS - CONTINUOUS COMPOUNDING

NOMINAL INTEREST RATE = 1.25 PERCENT

N	SINGLE-PAYMENT COMPOUND-AMOUNT FACTOR (F/P)	UNIFORM-SERIES COMPOUND-AMOUNT FACTOR (F/A)	UNIFORM-SERIES CAPITAL-RECOVERY FACTOR (A/P)	GRADIENT SERIES FACTOR (A/G)
1	1.0126	1.0000	1.01258	.0000
2	1.0253	2.0126	.50945	.4969
3	1.0382	3.0379	.34175	.9917
4	1.0513	4.0761	.25791	1.4844
5	1.0645	5.1274	.20761	1.9750
6	1.0779	6.1919	.17408	2.4635
7	1.0914	7.2698	.15013	2.9500
8	1.1052	8.3612	.13218	3.4344
9	1.1191	9.4664	.11822	3.9167
10	1.1331	10.5854	.10705	4.3969
11	1.1474	11.7186	.09791	4.8750
12	1.1618	12.8660	.09030	5.3511
13	1.1764	14.0278	.08387	5.8251
14	1.1912	15.2043	.07835	6.2970
15	1.2062	16.3955	.07357	6.7668
16	1.2214	17.6017	.06939	7.2346
17	1.2368	18.8232	.06570	7.7002
18	1.2523	20.0599	.06243	8.1638
19	1.2681	21.3122	.05950	8.6254
20	1.2840	22.5803	.05686	9.0848
21	1.3002	23.8643	.05448	9.5422
22	1.3165	25.1645	.05232	9.9975
23	1.3331	26.4810	.05034	10.4508
24	1.3499	27.8141	.04853	10.9019
25	1.3668	29.1640	.04687	11.3511
26	1.3840	30.5308	.04533	11.7981
27	1.4014	31.9149	.04391	12.2431
28	1.4191	33.3163	.04259	12.6860
29	1.4369	34.7354	.04137	13.1269
30	1.4550	36.1723	.04022	13.5657
31	1.4733	37.6273	.03915	14.0025
32	1.4918	39.1006	.03815	14.4372
33	1.5106	40.5924	.03721	14.8699
34	1.5296	42.1030	.03633	15.3005
35	1.5488	43.6326	.03550	15.7291
40	1.6487	51.5740	.03197	17.8413
45	1.7551	60.0276	.02924	19.9027
50	1.8682	69.0265	.02707	21.9137
55	1.9887	78.6057	.02530	23.8745
60	2.1170	88.8027	.02384	25.7857
65	2.2535	99.6573	.02261	27.6477
70	2.3989	111.2120	.02157	29.4608
75	2.5536	123.5120	.02067	31.2257
80	2.7183	136.6052	.01990	32.9429
85	2.8936	150.5429	.01922	34.6129
90	3.0802	165.3794	.01863	36.2363
95	3.2789	181.1728	.01810	37.8138
100	3.4903	197.9849	.01763	39.3459

COMPOUND INTEREST FACTORS - CONTINUOUS COMPOUNDING

NOMINAL INTEREST RATE = 1.50 PERCENT

N	SINGLE-PAYMENT COMPOUND-AMOUNT FACTOR (F/P)	UNIFORM-SERIES COMPOUND-AMOUNT FACTOR (F/A)	UNIFORM-SERIES CAPITAL-RECOVERY FACTOR (A/P)	GRADIENT SERIES FACTOR (A/G)
1	1.0151	1.0000	1.01511	.0000
2	1.0305	2.0151	.51136	.4963
3	1.0460	3.0456	.34346	.9900
4	1.0618	4.0916	.25952	1.4813
5	1.0779	5.1534	.20916	1.9700
6	1.0942	6.2313	.17559	2.4563
7	1.1107	7.3255	.15162	2.9400
8	1.1275	8.4362	.13365	3.4213
9	1.1445	9.5637	.11968	3.9000
10	1.1618	10.7082	.10850	4.3763
11	1.1794	11.8701	.09936	4.8501
12	1.1972	13.0495	.09174	5.3213
13	1.2153	14.2467	.08530	5.7901
14	1.2337	15.4620	.07979	6.2564
15	1.2523	16.6957	.07501	6.7202
16	1.2712	17.9480	.07083	7.1816
17	1.2905	19.2192	.06714	7.6404
18	1.3100	20.5097	.06387	8.0967
19	1.3298	21.8197	.06094	8.5506
20	1.3499	23.1494	.05831	9.0020
21	1.3703	24.4993	.05593	9.4509
22	1.3910	25.8695	.05377	9.8973
23	1.4120	27.2605	.05180	10.3413
24	1.4333	28.6725	.04999	10.7828
25	1.4550	30.1058	.04833	11.2218
26	1.4770	31.5608	.04680	11.6584
27	1.4993	33.0378	.04538	12.0925
28	1.5220	34.5371	.04407	12.5241
29	1.5450	36.0591	.04285	12.9533
30	1.5683	37.6040	.04171	13.3800
31	1.5920	39.1723	.04064	13.8043
32	1.6161	40.7644	.03964	14.2261
33	1.6405	42.3804	.03871	14.6455
34	1.6653	44.0209	.03783	15.0625
35	1.6905	45.6862	.03700	15.4770
40	1.8221	54.3979	.03350	17.5131
45	1.9640	63.7881	.03079	19.4890
50	2.1170	73.9096	.02864	21.4052
55	2.2819	84.8194	.02690	23.2622
60	2.4596	96.5789	.02547	25.0609
65	2.6512	109.2543	.02427	26.8018
70	2.8577	122.9169	.02325	28.4859
75	3.0802	137.6436	.02238	30.1140
80	3.3201	153.5173	.02163	31.6869
85	3.5787	170.6273	.02097	33.2056
90	3.8574	189.0699	.02040	34.6710
95	4.1579	208.9489	.01990	36.0842
100	4.4817	230.3761	.01945	37.4462

COMPOUND INTEREST FACTORS - CONTINUOUS COMPOUNDING

NOMINAL INTEREST RATE = 2.00 PERCENT

N	SINGLE-PAYMENT COMPOUND-AMOUNT FACTOR (F/P)	UNIFORM-SERIES COMPOUND-AMOUNT FACTOR (F/A)	UNIFORM-SERIES CAPITAL-RECOVERY FACTOR (A/P)	GRADIENT SERIES FACTOR (A/G)
1	1.0202	1.0000	1.02020	.0000
2	1.0408	2.0202	.51520	.4950
3	1.0010	3.0610	.34680	.9867
4	1.0833	4.1228	.26275	1.4750
5	1.1052	5.2061	.21228	1.9600
6	1.1275	6.3113	.17865	2.4417
7	1.1503	7.4388	.15463	2.9200
8	1.1735	8.5891	.13663	3.3950
9	1.1972	9.7626	.12263	3.8667
10	1.2214	10.9598	.11144	4.3351
11	1.2461	12.1812	.10230	4.8002
12	1.2712	13.4273	.09468	5.2619
13	1.2969	14.6985	.08824	5.7203
14	1.3231	15.9955	.08272	6.1754
15	1.3499	17.3186	.07794	6.6272
16	1.3771	18.6685	.07377	7.0757
17	1.4049	20.0456	.07009	7.5209
18	1.4333	21.4505	.06682	7.9628
19	1.4623	22.8839	.06390	8.4014
20	1.4918	24.3461	.06128	8.8368
21	1.5220	25.8380	.05890	9.2688
22	1.5527	27.3599	.05675	9.6976
23	1.5841	28.9126	.05479	10.1231
24	1.6161	30.4967	.05299	10.5453
25	1.6487	32.1128	.05134	10.9643
26	1.6820	33.7615	.04982	11.3800
27	1.7160	35.4435	.04842	11.7925
28	1.7507	37.1595	.04711	12.2018
29	1.7860	38.9102	.04590	12.6078
30	1.8221	40.6963	.04477	13.0106
31	1.8589	42.5184	.04372	13.4102
32	1.8965	44.3773	.04274	13.8065
33	1.9348	46.2738	.04181	14.1997
34	1.9739	48.2086	.04094	14.5897
35	2.0138	50.1824	.04013	14.9765
40	2.2255	60.6663	.03668	16.8630
45	2.4596	72.2528	.03404	18.6714
50	2.7183	85.0578	.03196	20.4028
55	3.0042	99.2096	.03028	22.0588
60	3.3201	114.8497	.02891	23.6409
65	3.6693	132.1346	.02777	25.1507
70	4.0552	151.2375	.02681	26.5899
75	4.4817	172.3494	.02600	27.9604
80	4.9530	195.6817	.02531	29.2640
85	5.4739	221.4679	.02472	30.5028
90	6.0496	249.9660	.02420	31.6786
95	6.6859	281.4613	.02375	32.7937
100	7.3891	316.2689	.02336	33.8499

COMPOUND INTEREST FACTORS - CONTINUOUS COMPOUNDING

NOMINAL INTEREST RATE = 3.00 PERCENT

N	SINGLE-PAYMENT COMPOUND-AMOUNT FACTOR (F/P)	UNIFORM-SERIES COMPOUND-AMOUNT FACTOR (F/A)	UNIFORM-SERIES CAPITAL-RECOVERY FACTOR (A/P)	GRADIENT SERIES FACTOR (A/G)
1	1.0305	1.0000	1.03045	.0000
2	1.0618	2.0305	.52296	.4925
3	1.0942	3.0923	.35384	.9800
4	1.1275	4.1865	.26932	1.4625
5	1.1618	5.3140	.21864	1.9400
6	1.1972	6.4758	.18488	2.4125
7	1.2337	7.6730	.16078	2.8801
8	1.2712	8.9067	.14273	3.3427
9	1.3100	10.1779	.12871	3.8002
10	1.3499	11.4879	.11750	4.2529
11	1.3910	12.8378	.10835	4.7005
12	1.4333	14.2287	.10073	5.1433
13	1.4770	15.6621	.09430	5.5811
14	1.5220	17.1390	.08880	6.0139
15	1.5683	18.6610	.08404	6.4419
16	1.6161	20.2293	.07989	6.8649
17	1.6653	21.8454	.07623	7.2831
18	1.7160	23.5107	.07299	7.6964
19	1.7683	25.2267	.07010	8.1048
20	1.8221	26.9950	.06750	8.5084
21	1.8776	28.8171	.06516	8.9072
22	1.9348	30.6947	.06303	9.3012
23	1.9937	32.6295	.06110	9.6904
24	2.0544	34.6232	.05934	10.0748
25	2.1170	36.6776	.05772	10.4545
26	2.1815	38.7946	.05623	10.8294
27	2.2479	40.9761	.05486	11.1996
28	2.3164	43.2240	.05359	11.5652
29	2.3869	45.5404	.05241	11.9261
30	2.4596	47.9273	.05132	12.2823
31	2.5345	50.3869	.05030	12.6339
32	2.6117	52.9214	.04935	12.9810
33	2.6912	55.5331	.04846	13.3235
34	2.7732	58.2243	.04763	13.6614
35	2.8577	60.9975	.04685	13.9948
40	3.3201	76.1830	.04358	15.5953
45	3.8574	93.8259	.04111	17.0874
50	4.4817	114.3242	.03920	18.4750
55	5.2070	138.1397	.03769	19.7623
60	6.0496	165.8094	.03649	20.9538
65	7.0287	197.9570	.03551	22.0541
70	8.1662	235.3072	.03470	23.0677
75	9.4877	278.7019	.03404	23.9996
80	11.0232	329.1193	.03349	24.8543
85	12.8071	387.6961	.03303	25.6368
90	14.8797	455.7526	.03265	26.3516
95	17.2878	534.8229	.03232	27.0032
100	20.0855	626.6895	.03205	27.5963

COMPOUND INTEREST FACTORS - CONTINUOUS COMPOUNDING

NOMINAL INTEREST RATE = 4.00 PERCENT

N	SINGLE-PAYMENT COMPOUND-AMOUNT FACTOR (F/P)	UNIFORM-SERIES COMPOUND-AMOUNT FACTOR (F/A)	UNIFORM-SERIES CAPITAL-RECOVERY FACTOR (A/P)	GRADIENT SERIES FACTOR (A/G)
1	1.0408	1.0000	1.04081	.0000
2	1.0833	2.0408	.53081	.4900
3	1.1275	3.1241	.36000	.9733
4	1.1735	4.2516	.27602	1.4500
5	1.2214	5.4251	.22514	1.9201
6	1.2712	6.6465	.19127	2.3834
7	1.3231	7.9178	.16711	2.8402
8	1.3771	9.2409	.14903	3.2904
9	1.4333	10.6180	.13499	3.7339
10	1.4918	12.0513	.12379	4.1709
11	1.5527	13.5432	.11465	4.6013
12	1.6161	15.0959	.10705	5.0252
13	1.6820	16.7120	.10065	5.4425
14	1.7507	18.3940	.09518	5.8534
15	1.8221	20.1447	.09045	6.2578
16	1.8965	21.9668	.08633	6.6558
17	1.9739	23.8633	.08272	7.0473
18	2.0544	25.8371	.07951	7.4326
19	2.1383	27.8916	.07666	7.8114
20	2.2255	30.0298	.07411	8.1840
21	2.3164	32.2554	.07181	8.5503
22	2.4109	34.5717	.06974	8.9104
23	2.5093	36.9826	.06785	9.2644
24	2.6117	39.4919	.06613	9.6122
25	2.7183	42.1036	.06456	9.9539
26	2.8292	44.8219	.06312	10.2896
27	2.9447	47.6511	.06180	10.6193
28	3.0649	50.5958	.06058	10.9431
29	3.1899	53.6607	.05945	11.2609
30	3.3201	56.8506	.05840	11.5730
31	3.4556	60.1707	.05743	11.8792
32	3.5966	63.6263	.05653	12.1797
33	3.7434	67.2230	.05569	12.4746
34	3.8962	70.9664	.05490	12.7638
35	4.0552	74.8626	.05417	13.0475
40	4.9530	96.8625	.05113	14.3845
45	6.0496	123.7332	.04889	15.5918
50	7.3891	156.5532	.04720	16.6775
55	9.0250	196.6396	.04590	17.6498
60	11.0232	245.6012	.04488	18.5172
65	13.4637	305.4031	.04409	19.2882
70	16.4446	378.4453	.04345	19.9710
75	20.0855	467.6593	.04295	20.5737
80	24.5325	576.6254	.04255	21.1038
85	29.9641	709.7170	.04222	21.5687
90	36.5982	872.2754	.04196	21.9751
95	44.7012	1070.8247	.04174	22.3295
100	54.5982	1313.3333	.04157	22.6376

COMPOUND INTEREST FACTORS - CONTINUOUS COMPOUNDING

NOMINAL INTEREST RATE = 5.00 PERCENT

N	SINGLE-PAYMENT COMPOUND-AMOUNT FACTOR (F/P)	UNIFORM-SERIES COMPOUND-AMOUNT FACTOR (F/A)	UNIFORM-SERIES CAPITAL-RECOVERY FACTOR (A/P)	GRADIENT SERIES FACTOR (A/G)
1	1.0513	1.0000	1.05127	.0000
2	1.1052	2.0513	.53877	.4875
3	1.1618	3.1564	.36808	.9667
4	1.2214	4.3183	.28284	1.4375
5	1.2840	5.5397	.23179	1.9001
6	1.3499	6.8237	.19782	2.3544
7	1.4191	8.1736	.17362	2.8004
8	1.4918	9.5926	.15552	3.2382
9	1.5683	11.0845	.14149	3.6678
10	1.6487	12.6528	.13031	4.0892
11	1.7333	14.3015	.12119	4.5025
12	1.8221	16.0347	.11364	4.9077
13	1.9155	17.8569	.10727	5.3049
14	2.0138	19.7724	.10185	5.6941
15	2.1170	21.7862	.09717	6.0753
16	2.2255	23.9032	.09311	6.4487
17	2.3396	26.1287	.08954	6.8143
18	2.4596	28.4683	.08640	7.1720
19	2.5857	30.9279	.08360	7.5221
20	2.7183	33.5137	.08111	7.8646
21	2.8577	36.2319	.07887	8.1996
22	3.0042	39.0896	.07685	8.5270
23	3.1582	42.0938	.07503	8.8471
24	3.3201	45.2519	.07337	9.1599
25	3.4903	48.5721	.07186	9.4654
26	3.6693	52.0624	.07048	9.7638
27	3.8574	55.7317	.06921	10.0551
28	4.0552	59.5891	.06805	10.3395
29	4.2631	63.6443	.06698	10.6170
30	4.4817	67.9074	.06600	10.8877
31	4.7115	72.3891	.06509	11.1517
32	4.9530	77.1006	.06424	11.4091
33	5.2070	82.0536	.06346	11.6601
34	5.4739	87.2606	.06273	11.9046
35	5.7546	92.7346	.06205	12.1429
40	7.3891	124.6132	.05930	13.2435
45	9.4877	165.5462	.05731	14.2024
50	12.1825	218.1052	.05586	15.0329
55	15.6426	285.5923	.05477	15.7480
60	20.0855	372.2475	.05396	16.3604
65	25.7903	483.5149	.05334	16.8822
70	33.1155	626.3851	.05287	17.3245
75	42.5211	809.8341	.05251	17.6979
80	54.5982	1045.3872	.05223	18.0116
85	70.1054	1347.8435	.05201	18.2742
90	90.0171	1736.2049	.05185	18.4931
95	115.5843	2234.8710	.05172	18.6751
100	148.4132	2875.1708	.05162	18.8258

COMPOUND INTEREST FACTORS - CONTINUOUS COMPOUNDING

NOMINAL INTEREST RATE = 6.00 PERCENT

N	SINGLE-PAYMENT COMPOUND-AMOUNT FACTOR (F/P)	UNIFORM-SERIES COMPOUND-AMOUNT FACTOR (F/A)	UNIFORM-SERIES CAPITAL-RECOVERY FACTOR (A/P)	GRADIENT SERIES FACTOR (A/G)
1	1.0618	1.0000	1.06184	.0000
2	1.1275	2.0618	.54684	.4850
3	1.1972	3.1893	.37538	.9600
4	1.2712	4.3866	.28981	1.4251
5	1.3499	5.6578	.23858	1.8802
6	1.4333	7.0077	.20454	2.3254
7	1.5220	8.4410	.18031	2.7607
8	1.6161	9.9629	.16221	3.1862
9	1.7160	11.5790	.14820	3.6020
10	1.8221	13.2950	.13705	4.0080
11	1.9348	15.1171	.12799	4.4043
12	2.0544	17.0519	.12048	4.7911
13	2.1815	19.1064	.11418	5.1684
14	2.3164	21.2878	.10881	5.5363
15	2.4596	23.6042	.10420	5.8949
16	2.6117	26.0638	.10020	6.2442
17	2.7732	28.6755	.09671	6.5845
18	2.9447	31.4487	.09363	6.9156
19	3.1268	34.3934	.09091	7.2379
20	3.3201	37.5202	.08849	7.5514
21	3.5254	40.8403	.08632	7.8562
22	3.7434	44.3657	.08438	8.1525
23	3.9749	48.1091	.08262	8.4403
24	4.2207	52.0840	.08104	8.7199
25	4.4817	56.3047	.07960	8.9912
26	4.7588	60.7864	.07829	9.2546
27	5.0531	65.5452	.07709	9.5101
28	5.3656	70.5983	.07600	9.7578
29	5.6973	75.9639	.07500	9.9980
30	6.0496	81.6612	.07408	10.2307
31	6.4237	87.7109	.07324	10.4560
32	6.8210	94.1346	.07246	10.6743
33	7.2427	100.9556	.07174	10.8855
34	7.6906	108.1983	.07108	11.0899
35	8.1662	115.8889	.07047	11.2876
40	11.0232	162.0915	.06801	12.1809
45	14.8797	224.4584	.06629	12.9295
50	20.0855	308.6449	.06508	13.5519
55	27.1126	422.2849	.06420	14.0654
60	36.5982	575.6828	.06357	14.4862
65	49.4024	782.7483	.06311	14.8288
70	66.6863	1062.2574	.06278	15.1060
75	90.0171	1439.5553	.06253	15.3291
80	121.5104	1948.8543	.06235	15.5078
85	164.0219	2636.3359	.06222	15.6503
90	221.4064	3564.3390	.06212	15.7633
95	298.8674	4817.0122	.06204	15.8527
100	403.4288	6507.9442	.06199	15.9232

COMPOUND INTEREST FACTORS - CONTINUOUS COMPOUNDING

NOMINAL INTEREST RATE = 7.00 PERCENT

N	SINGLE-PAYMENT COMPOUND-AMOUNT FACTOR (F/P)	UNIFORM-SERIES COMPOUND-AMOUNT FACTOR (F/A)	UNIFORM-SERIES CAPITAL-RECOVERY FACTOR (A/P)	GRADIENT SERIES FACTOR (A/G)
1	1.0725	1.0000	1.07251	.0000
2	1.1503	2.0725	.55502	.4825
3	1.2337	3.2228	.38280	.9534
4	1.3231	4.4565	.29690	1.4126
5	1.4191	5.7796	.24553	1.8603
6	1.5220	7.1987	.21142	2.2964
7	1.6323	8.7206	.18718	2.7211
8	1.7507	10.3529	.16910	3.1344
9	1.8776	12.1036	.15513	3.5364
10	2.0138	13.9812	.14403	3.9272
11	2.1598	15.9950	.13503	4.3069
12	2.3164	18.1547	.12759	4.6755
13	2.4843	20.4711	.12136	5.0333
14	2.6645	22.9554	.11607	5.3804
15	2.8577	25.6199	.11154	5.7168
16	3.0649	28.4775	.10762	6.0428
17	3.2871	31.5424	.10421	6.3585
18	3.5254	34.8295	.10122	6.6640
19	3.7810	38.3549	.09858	6.9596
20	4.0552	42.1359	.09624	7.2453
21	4.3492	46.1911	.09416	7.5215
22	4.6646	50.5404	.09229	7.7881
23	5.0028	55.2050	.09062	8.0456
24	5.3656	60.2078	.08912	8.2940
25	5.7546	65.5733	.08776	8.5335
26	6.1719	71.3279	.08653	8.7643
27	6.6194	77.4998	.08541	8.9867
28	7.0993	84.1192	.08440	9.2009
29	7.6141	91.2185	.08347	9.4070
30	8.1662	98.8326	.08263	9.6052
31	8.7583	106.9987	.08185	9.7958
32	9.3933	115.7570	.08115	9.9790
33	10.0744	125.1504	.08050	10.1550
34	10.8049	135.2248	.07990	10.3239
35	11.5883	146.0297	.07936	10.4860
40	16.4446	213.0056	.07720	11.2017
45	23.3361	308.0489	.07575	11.7769
50	33.1155	442.9218	.07477	12.2347
55	46.9931	634.3155	.07408	12.5957
60	66.6863	905.9161	.07361	12.8781
65	94.6324	1291.3358	.07328	13.0973
70	134.2898	1838.2723	.07305	13.2664
75	190.5663	2614.4121	.07289	13.3959
80	270.4264	3715.8070	.07278	13.4946
85	383.7533	5278.7607	.07270	13.5695
90	544.5719	7496.6976	.07264	13.6260
95	772.7843	10644.0999	.07260	13.6685
100	1096.6332	15110.4764	.07257	13.7003

COMPOUND INTEREST FACTORS - CONTINUOUS COMPOUNDING

NOMINAL INTEREST RATE = 8.00 PERCENT

N	SINGLE-PAYMENT COMPOUND-AMOUNT FACTOR (F/P)	UNIFORM-SERIES COMPOUND-AMOUNT FACTOR (F/A)	UNIFORM-SERIES CAPITAL-RECOVERY FACTOR (A/P)	GRADIENT SERIES FACTOR (A/G)
1	1.0833	1.0000	1.08329	.0000
2	1.1735	2.0833	.56330	.4800
3	1.2712	3.2568	.39034	.9467
4	1.3771	4.5280	.30413	1.4002
5	1.4918	5.9052	.25263	1.8404
6	1.6161	7.3970	.21848	2.2676
7	1.7507	9.0131	.19424	2.6817
8	1.8965	10.7637	.17619	3.0829
9	2.0544	12.6602	.16227	3.4713
10	2.2255	14.7147	.15125	3.8470
11	2.4109	16.9402	.14232	4.2102
12	2.6117	19.3511	.13496	4.5611
13	2.8292	21.9628	.12882	4.8998
14	3.0649	24.7920	.12362	5.2265
15	3.3201	27.8569	.11918	5.5415
16	3.5966	31.1770	.11536	5.8449
17	3.8962	34.7736	.11204	6.1369
18	4.2207	38.6698	.10915	6.4178
19	4.5722	42.8905	.10660	6.6879
20	4.9530	47.4627	.10436	6.9473
21	5.3656	52.4158	.10237	7.1963
22	5.8124	57.7813	.10059	7.4352
23	6.2965	63.5938	.09901	7.6642
24	6.8210	69.8903	.09760	7.8836
25	7.3891	76.7113	.09632	8.0937
26	8.0045	84.1003	.09518	8.2947
27	8.6711	92.1048	.09414	8.4870
28	9.3933	100.7759	.09321	8.6707
29	10.1757	110.1693	.09236	8.8461
30	11.0232	120.3449	.09160	9.0136
31	11.9413	131.3681	.09090	9.1734
32	12.9358	143.3094	.09026	9.3257
33	14.0132	156.2452	.08969	9.4708
34	15.1803	170.2584	.08916	9.6090
35	16.4446	185.4387	.08868	9.7405
40	24.5325	282.5472	.08683	10.3069
45	36.5982	427.4161	.08563	10.7426
50	54.5982	643.5351	.08484	11.0738
55	81.4509	965.9467	.08432	11.3230
60	121.5104	1446.9283	.08398	11.5088
65	181.2722	2164.4686	.08375	11.6461
70	270.4264	3234.9129	.08360	11.7469
75	403.4288	4831.8281	.08349	11.8203
80	601.8450	7214.1457	.08343	11.8735
85	897.8473	10768.1458	.08338	11.9119
90	1339.4308	16070.0911	.08335	11.9394
95	1998.1959	23979.6640	.08333	11.9591
100	2980.9580	35779.3601	.08332	11.9731

COMPOUND INTEREST FACTORS - CONTINUOUS COMPOUNDING

NOMINAL INTEREST RATE = 9.00 PERCENT

N	SINGLE-PAYMENT COMPOUND-AMOUNT FACTOR (F/P)	UNIFORM-SERIES COMPOUND-AMOUNT FACTOR (F/A)	UNIFORM-SERIES CAPITAL-RECOVERY FACTOR (A/P)	GRADIENT SERIES FACTOR (A/G)
1	1.0942	1.0000	1.09417	.0000
2	1.1972	2.0942	.57169	.4775
3	1.3100	3.2914	.39800	.9401
4	1.4333	4.6014	.31150	1.3878
5	1.5683	6.0347	.25988	1.8206
6	1.7160	7.6030	.22570	2.2388
7	1.8776	9.3190	.20148	2.6424
8	2.0544	11.1966	.18349	3.0316
9	2.2479	13.2510	.16964	3.4065
10	2.4596	15.4990	.15869	3.7674
11	2.6912	17.9586	.14986	4.1145
12	2.9447	20.6498	.14260	4.4479
13	3.2220	23.5945	.13656	4.7680
14	3.5254	26.8165	.13146	5.0750
15	3.8574	30.3419	.12713	5.3691
16	4.2207	34.1993	.12341	5.6507
17	4.6182	38.4200	.12020	5.9201
18	5.0531	43.0382	.11741	6.1776
19	5.5290	48.0913	.11497	6.4234
20	6.0496	53.6202	.11282	6.6579
21	6.6194	59.6699	.11093	6.8815
22	7.2427	66.2893	.10926	7.0945
23	7.9248	73.5320	.10777	7.2972
24	8.6711	81.4568	.10645	7.4900
25	9.4877	90.1280	.10527	7.6732
26	10.3812	99.6157	.10421	7.8471
27	11.3589	109.9969	.10327	8.0122
28	12.4286	121.3558	.10241	8.1686
29	13.5991	133.7844	.10165	8.3168
30	14.8797	147.3835	.10096	8.4572
31	16.2810	162.2632	.10034	8.5899
32	17.8143	178.5442	.09978	8.7155
33	19.4919	196.3585	.09927	8.8340
34	21.3276	215.8504	.09881	8.9460
35	23.3361	237.1780	.09839	9.0516
40	36.5982	378.0038	.09682	9.4950
45	57.3975	598.8626	.09584	9.8207
50	90.0171	945.2382	.09523	10.0569
55	141.1750	1488.4633	.09485	10.2262
60	221.4064	2340.4098	.09460	10.3464
65	347.2344	3676.5279	.09445	10.4309
70	544.5719	5771.9782	.09435	10.4898
75	854.0588	9058.2984	.09428	10.5307
80	1339.4308	14212.2744	.09424	10.5588
85	2100.6456	22295.3179	.09422	10.5781
90	3294.4681	34972.0534	.09420	10.5913
95	5166.7544	54853.1321	.09419	10.6002
100	8103.0839	86032.8702	.09419	10.6063

COMPOUND INTEREST FACTORS - CONTINUOUS COMPOUNDING

NOMINAL INTEREST RATE = 10.00 PERCENT

N	SINGLE-PAYMENT COMPOUND-AMOUNT FACTOR (F/P)	UNIFORM-SERIES COMPOUND-AMOUNT FACTOR (F/A)	UNIFORM-SERIES CAPITAL-RECOVERY FACTOR (A/P)	GRADIENT SERIES FACTOR (A/G)
1	1.1052	1.0000	1.10517	.0000
2	1.2214	2.1052	.58019	.4750
3	1.3499	0.0200	.40578	.9334
4	1.4918	4.6764	.31901	1.3754
5	1.6487	6.1683	.26729	1.8009
6	1.8221	7.8170	.23310	2.2101
7	2.0138	9.6391	.20892	2.6033
8	2.2255	11.6528	.19099	2.9806
9	2.4596	13.8784	.17723	3.3423
10	2.7183	16.3380	.16638	3.6886
11	3.0042	19.0563	.15765	4.0198
12	3.3201	22.0604	.15050	4.3362
13	3.6693	25.3806	.14457	4.6381
14	4.0552	29.0499	.13959	4.9260
15	4.4817	33.1051	.13538	5.2001
16	4.9530	37.5867	.13178	5.4608
17	5.4739	42.5398	.12868	5.7086
18	6.0496	48.0137	.12600	5.9437
19	6.6859	54.0634	.12367	6.1667
20	7.3891	60.7493	.12163	6.3780
21	8.1662	68.1383	.11985	6.5779
22	9.0250	76.3045	.11828	6.7669
23	9.9742	85.3295	.11689	6.9454
24	11.0232	95.3037	.11566	7.1139
25	12.1825	106.3269	.11458	7.2727
26	13.4637	118.5094	.11361	7.4223
27	14.8797	131.9731	.11275	7.5630
28	16.4446	146.8528	.11198	7.6954
29	18.1741	163.2975	.11129	7.8197
30	20.0855	181.4716	.11068	7.9365
31	22.1980	201.5572	.11013	8.0459
32	24.5325	223.7551	.10964	8.1485
33	27.1126	248.2876	.10920	8.2446
34	29.9641	275.4003	.10880	8.3345
35	33.1155	305.3644	.10845	8.4185
40	54.5982	509.6290	.10713	8.7620
45	90.0171	846.4044	.10635	9.0028
50	148.4132	1401.6532	.10588	9.1691
55	244.6919	2317.1038	.10560	9.2826
60	403.4288	3826.4266	.10543	9.3592
65	665.1416	6314.8791	.10533	9.4105
70	1096.6332	10417.6438	.10527	9.4444
75	1808.0424	17181.9591	.10523	9.4668
80	2980.9580	28334.4297	.10521	9.4815
85	4914.7688	46721.7452	.10519	9.4910
90	8103.0839	77037.3034	.10518	9.4972
95	13359.7268	127019.2091	.10518	9.5012
100	22026.4658	209425.4400	.10518	9.5038

COMPOUND INTEREST FACTORS - CONTINUOUS COMPOUNDING

NOMINAL INTEREST RATE = 12.00 PERCENT

N	SINGLE-PAYMENT COMPOUND-AMOUNT FACTOR (F/P)	UNIFORM-SERIES COMPOUND-AMOUNT FACTOR (F/A)	UNIFORM-SERIES CAPITAL-RECOVERY FACTOR (A/P)	GRADIENT SERIES FACTOR (A/G)
1	1.1275	1.0000	1.12750	.0000
2	1.2712	2.1275	.59753	.4700
3	1.4333	3.3987	.42172	.9202
4	1.6161	4.8321	.33445	1.3506
5	1.8221	6.4481	.28258	1.7615
6	2.0544	8.2703	.24841	2.1531
7	2.3164	10.3247	.22435	2.5257
8	2.6117	12.6411	.20660	2.8796
9	2.9447	15.2528	.19306	3.2153
10	3.3201	18.1974	.18245	3.5332
11	3.7434	21.5176	.17397	3.8337
12	4.2207	25.2610	.16708	4.1174
13	4.7588	29.4817	.16142	4.3848
14	5.3656	34.2405	.15670	4.6364
15	6.0496	39.6061	.15275	4.8728
16	6.8210	45.6557	.14940	5.0946
17	7.6906	52.4767	.14655	5.3025
18	8.6711	60.1673	.14412	5.4969
19	9.7767	68.8384	.14202	5.6785
20	11.0232	78.6151	.14022	5.8480
21	12.4286	89.6383	.13865	6.0058
22	14.0132	102.0669	.13729	6.1527
23	15.7998	116.0801	.13611	6.2893
24	17.8143	131.8799	.13508	6.4160
25	20.0855	149.6942	.13418	6.5334
26	22.6464	169.7797	.13339	6.6422
27	25.5337	192.4261	.13269	6.7428
28	28.7892	217.9598	.13208	6.8357
29	32.4597	246.7490	.13155	6.9215
30	36.5982	279.2087	.13108	7.0006
31	41.2644	315.8070	.13066	7.0734
32	46.5255	357.0714	.13030	7.1404
33	52.4573	403.5968	.12997	7.2020
34	59.1455	456.0542	.12969	7.2586
35	66.6863	515.1996	.12944	7.3105
40	121.5104	945.2031	.12855	7.5114
45	221.4064	1728.7205	.12808	7.6392
50	403.4288	3156.3822	.12781	7.7191

COMPOUND INTEREST FACTORS - CONTINUOUS COMPOUNDING

NOMINAL INTEREST RATE = 15.00 PERCENT

N	SINGLE-PAYMENT COMPOUND-AMOUNT FACTOR (F/P)	UNIFORM-SERIES COMPOUND-AMOUNT FACTOR (F/A)	UNIFORM-SERIES CAPITAL-RECOVERY FACTOR (A/P)	GRADIENT SERIES FACTOR (A/G)
1	1.1618	1.0000	1.16183	.0000
2	1.3499	2.1618	.62440	.4626
3	1.5683	3.5117	.44660	.9004
4	1.8221	5.0800	.35868	1.3137
5	2.1170	6.9021	.30672	1.7029
6	2.4596	9.0191	.27271	2.0685
7	2.8577	11.4787	.24895	2.4110
8	3.3201	14.3364	.23159	2.7311
9	3.8574	17.6565	.21847	3.0295
10	4.4817	21.5139	.20832	3.3070
11	5.2070	25.9956	.20030	3.5645
12	6.0496	31.2026	.19388	3.8028
13	7.0287	37.2522	.18868	4.0228
14	8.1662	44.2809	.18442	4.2255
15	9.4877	52.4471	.18090	4.4119
16	11.0232	61.9348	.17798	4.5829
17	12.8071	72.9580	.17554	4.7394
18	14.8797	85.7651	.17349	4.8823
19	17.2878	100.6448	.17177	5.0126
20	20.0855	117.9326	.17031	5.1312
21	23.3361	138.0182	.16908	5.2390
22	27.1126	161.3542	.16803	5.3367
23	31.5004	188.4669	.16714	5.4251
24	36.5982	219.9673	.16638	5.5050
25	42.5211	256.5655	.16573	5.5771
26	49.4024	299.0866	.16518	5.6420
27	57.3975	348.4890	.16470	5.7004
28	66.6863	405.8865	.16430	5.7529
29	77.4785	472.5728	.16395	5.8000
30	90.0171	550.0513	.16365	5.8421
31	104.5850	640.0684	.16340	5.8799
32	121.5104	744.6534	.16318	5.9136
33	141.1750	866.1638	.16299	5.9437
34	164.0219	1007.3388	.16283	5.9706
35	190.5663	1171.3607	.16269	5.9945
40	403.4288	2486.6727	.16224	6.0798
45	854.0588	5271.1883	.16202	6.1264
50	1808.0424	11166.0078	.16192	6.1515

COMPOUND INTEREST FACTORS - CONTINUOUS COMPOUNDING

NOMINAL INTEREST RATE = 20.00 PERCENT

N	SINGLE-PAYMENT COMPOUND-AMOUNT FACTOR (F/P)	UNIFORM-SERIES COMPOUND-AMOUNT FACTOR (F/A)	UNIFORM-SERIES CAPITAL-RECOVERY FACTOR (A/P)	GRADIENT SERIES FACTOR (A/G)
1	1.2214	1.0000	1.22140	.0000
2	1.4918	2.2214	.67157	.4502
3	1.8221	3.7132	.49071	.8675
4	2.2255	5.5353	.40206	1.2528
5	2.7183	7.7609	.35025	1.6068
6	3.3201	10.4792	.31683	1.9306
7	4.0552	13.7993	.29387	2.2255
8	4.9530	17.8545	.27741	2.4929
9	6.0496	22.8075	.26525	2.7344
10	7.3891	28.8572	.25606	2.9515
11	9.0250	36.2462	.24899	3.1459
12	11.0232	45.2712	.24349	3.3194
13	13.4637	56.2944	.23917	3.4736
14	16.4446	69.7581	.23574	3.6102
15	20.0855	86.2028	.23300	3.7307
16	24.5325	106.2883	.23081	3.8367
17	29.9641	130.8209	.22905	3.9297
18	36.5982	160.7850	.22762	4.0110
19	44.7012	197.3832	.22647	4.0819
20	54.5982	242.0844	.22553	4.1435
21	66.6863	296.6825	.22477	4.1970
22	81.4509	363.3689	.22415	4.2432
23	99.4843	444.8197	.22365	4.2831
24	121.5104	544.3040	.22324	4.3175
25	148.4132	665.8145	.22290	4.3471
26	181.2722	814.2276	.22263	4.3724
27	221.4064	995.4999	.22241	4.3942
28	270.4264	1216.9063	.22222	4.4127
29	330.2996	1487.3327	.22208	4.4286
30	403.4288	1817.6323	.22195	4.4421
31	492.7490	2221.0610	.22185	4.4536
32	601.8450	2713.8101	.22177	4.4634
33	735.0952	3315.6551	.22170	4.4717
34	897.8473	4050.7503	.22165	4.4787
35	1096.6332	4948.5976	.22160	4.4847
40	2980.9580	13459.4438	.22148	4.5032
45	8103.0839	36594.3225	.22143	4.5111
50	22026.4658	99481.4427	.22141	4.5144

COMPOUND INTEREST FACTORS - CONTINUOUS COMPOUNDING

NOMINAL INTEREST RATE = 25.00 PERCENT

N	SINGLE-PAYMENT COMPOUND-AMOUNT FACTOR (F/P)	UNIFORM-SERIES COMPOUND-AMOUNT FACTOR (F/A)	UNIFORM-SERIES CAPITAL-RECOVERY FACTOR (A/P)	GRADIENT SERIES FACTOR (A/G)
1	1.2840	1.0000	1.28403	.0000
2	1.6487	2.2840	.72185	.4378
3	2.1170	3.9327	.53830	.8350
4	2.7183	6.0497	.44932	1.1929
5	3.4903	8.7680	.39808	1.5131
6	4.4817	12.2584	.36560	1.7975
7	5.7546	16.7401	.34376	2.0486
8	7.3891	22.4947	.32848	2.2687
9	9.4877	29.8837	.31749	2.4605
10	12.1825	39.3715	.30942	2.6266
11	15.6426	51.5539	.30342	2.7696
12	20.0855	67.1966	.29891	2.8921
13	25.7903	87.2821	.29548	2.9964
14	33.1155	113.0725	.29287	3.0849
15	42.5211	146.1879	.29087	3.1595
16	54.5982	188.7090	.28932	3.2223
17	70.1054	243.3071	.28814	3.2748
18	90.0171	313.4126	.28722	3.3186
19	115.5843	403.4297	.28650	3.3550
20	148.4132	519.0140	.28595	3.3851
21	190.5663	667.4271	.28552	3.4100
22	244.6919	857.9934	.28519	3.4305
23	314.1907	1102.6853	.28493	3.4474
24	403.4288	1416.8760	.28473	3.4612
25	518.0128	1820.3048	.28457	3.4725
26	665.1416	2338.3176	.28445	3.4817
27	854.0588	3003.4592	.28436	3.4892
28	1096.6332	3857.5180	.28428	3.4953
29	1408.1048	4954.1512	.28423	3.5002
30	1808.0424	6362.2560	.28418	3.5042
31	2321.5724	8170.2984	.28415	3.5075
32	2980.9580	10491.8708	.28412	3.5101
33	3827.6258	13472.8288	.28410	3.5122
34	4914.7688	17300.4546	.28408	3.5139
35	6310.6881	22215.2235	.28407	3.5153

COMPOUND INTEREST FACTORS - CONTINUOUS COMPOUNDING

NOMINAL INTEREST RATE = 30.00 PERCENT

N	SINGLE-PAYMENT COMPOUND-AMOUNT FACTOR (F/P)	UNIFORM-SERIES COMPOUND-AMOUNT FACTOR (F/A)	UNIFORM-SERIES CAPITAL-RECOVERY FACTOR (A/P)	GRADIENT SERIES FACTOR (A/G)
1	1.3499	1.0000	1.34986	.0000
2	1.8221	2.3499	.77542	.4256
3	2.4596	4.1720	.58955	.8029
4	3.3201	6.6316	.50065	1.1342
5	4.4817	9.9517	.45034	1.4222
6	6.0496	14.4334	.41914	1.6701
7	8.1662	20.4830	.39868	1.8815
8	11.0232	28.6492	.38476	2.0601
9	14.8797	39.6724	.37507	2.2099
10	20.0855	54.5521	.36819	2.3343
11	27.1126	74.6376	.36326	2.4370
12	36.5982	101.7503	.35969	2.5212
13	49.4024	138.3485	.35709	2.5897
14	66.6863	187.7510	.35519	2.6452
15	90.0171	254.4373	.35379	2.6898
16	121.5104	344.4544	.35276	2.7255
17	164.0219	465.9649	.35200	2.7540
18	221.4064	629.9868	.35145	2.7766
19	298.8674	851.3932	.35103	2.7945
20	403.4288	1150.2606	.35073	2.8086
21	544.5719	1553.6894	.35050	2.8197
22	735.0952	2098.2613	.35034	2.8283
23	992.2747	2833.3565	.35021	2.8351
24	1339.4308	3825.6312	.35012	2.8404
25	1808.0424	5165.0619	.35005	2.8445
26	2440.6020	6973.1044	.35000	2.8476
27	3294.4681	9413.7063	.34997	2.8501
28	4447.0667	12708.1744	.34994	2.8520
29	6002.9122	17155.2412	.34992	2.8535
30	8103.0839	23158.1534	.34990	2.8546
31	10938.0192	31261.2373	.34989	2.8555
32	14764.7816	42199.2565	.34988	2.8561
33	19930.3704	56964.0381	.34988	2.8566
34	26903.1861	76894.4085	.34987	2.8570
35	36315.5027	103797.5946	.34987	2.8573

COMPOUND INTEREST FACTORS - CONTINUOUS COMPOUNDING

NOMINAL INTEREST RATE = 40.00 PERCENT

N	SINGLE-PAYMENT COMPOUND-AMOUNT FACTOR (F/P)	UNIFORM-SERIES COMPOUND-AMOUNT FACTOR (F/A)	UNIFORM-SERIES CAPITAL-RECOVERY FACTOR (A/P)	GRADIENT SERIES FACTOR (A/G)
1	1.4918	1.0000	1.49182	.0000
2	2.2255	2.4918	.89314	.4013
3	3.3201	4.7174	.70381	.7402
4	4.9530	8.0375	.61624	1.0214
5	7.3891	12.9905	.56880	1.2507
6	11.0232	20.3796	.54089	1.4346
7	16.4446	31.4027	.52367	1.5800
8	24.5325	47.8474	.51272	1.6933
9	36.5982	72.3799	.50564	1.7804
10	54.5982	108.9782	.50100	1.8467
11	81.4509	163.5763	.49794	1.8965
12	121.5104	245.0272	.49591	1.9337
13	181.2722	366.5376	.49455	1.9611
14	270.4264	547.8098	.49365	1.9813
15	403.4288	818.2362	.49305	1.9960
16	601.8450	1221.6650	.49264	2.0066
17	897.8473	1823.5101	.49237	2.0143
18	1339.4308	2721.3574	.49219	2.0198
19	1998.1959	4060.7881	.49207	2.0237
20	2980.9580	6058.9840	.49199	2.0265
21	4447.0667	9039.9420	.49194	2.0285
22	6634.2440	13487.0088	.49190	2.0299
23	9897.1291	20121.2528	.49187	2.0309
24	14764.7816	30018.3818	.49186	2.0316
25	22026.4658	44783.1634	.49185	2.0321
26	32859.6257	66809.6292	.49184	2.0325
27	49020.8011	99669.2549	.49183	2.0327
28	73130.4418	148690.0560	.49183	2.0329
29	**********	221820.4978	.49183	2.0330
30	**********	330918.2971	.49183	2.0331

COMPOUND INTEREST FACTORS - CONTINUOUS COMPOUNDING

NOMINAL INTEREST RATE = 50.00 PERCENT

N	SINGLE-PAYMENT COMPOUND-AMOUNT FACTOR (F/P)	UNIFORM-SERIES COMPOUND-AMOUNT FACTOR (F/A)	UNIFORM-SERIES CAPITAL-RECOVERY FACTOR (A/P)	GRADIENT SERIES FACTOR (A/G)
1	1.6487	1.0000	1.64872	.0000
2	2.7183	2.6487	1.02626	.3775
3	4.4817	5.3670	.83504	.6798
4	7.3891	9.8487	.75026	.9154
5	12.1825	17.2377	.70673	1.0944
6	20.0855	29.4202	.68271	1.2271
7	33.1155	49.5058	.66892	1.3235
8	54.5982	82.6212	.66082	1.3922
9	90.0171	137.2194	.65601	1.4404
10	148.4132	227.2365	.65312	1.4737
11	244.6919	375.6497	.65138	1.4964
12	403.4288	620.3416	.65033	1.5117
13	665.1416	1023.7704	.64970	1.5219
14	1096.6332	1688.9120	.64931	1.5287
15	1808.0424	2785.5452	.64908	1.5332
16	2980.9580	4593.5876	.64894	1.5361
17	4914.7688	7574.5456	.64885	1.5380
18	8103.0839	12489.3144	.64880	1.5393
19	13359.7268	20592.3984	.64877	1.5401
20	22026.4658	33952.1252	.64875	1.5406
21	36315.5027	55978.5910	.64874	1.5409
22	59874.1417	92294.0937	.64873	1.5411
23	98715.7710	152168.2354	.64873	1.5413
24	**********	250884.0064	.64873	1.5413
25	**********	413638.7978	.64872	1.5414

Appendix D

Annual versus Continuous Uniform Payment Factors

NOMINAL INTEREST RATE, %	RATIO OF ANNUAL TO CONTINUOUS UNIFORM PAYMENTS ($A/\bar{A}$)
1	1.005017
2	1.010067
3	1.015151
4	1.020269
5	1.025422
6	1.030609
7	1.035831
8	1.041088
9	1.046381
10	1.051709
11	1.057073
12	1.062474
13	1.067911
14	1.073384
15	1.078895
16	1.084443
17	1.090029
18	1.095652
19	1.101314
20	1.107014
21	1.112753
22	1.118531
23	1.124348
24	1.130205
25	1.136102
26	1.142039
27	1.148016
28	1.154035
29	1.160095
30	1.166196
31	1.172339
32	1.178524
33	1.184752
34	1.191022
35	1.197336
36	1.203693
37	1.210094
38	1.216538
39	1.223028
40	1.229562
41	1.236141
42	1.242766
43	1.249436
44	1.256153
45	1.262916
46	1.269726
47	1.276583
48	1.283488
49	1.290441
50	1.297443

Index